STATISTICS FOR HORTICULTURAL SCIENCE

THE AUTHORS

Dr. K. Umakrishna is working as Professor in Statistics at Dr. YSRHU- College of Horticulture, Venkataramannagudem. He has more than 32 years of teaching experience, published one text book, 70 research papers in National & International Journals and 30 abstracts in conferences. He has guided 20 Ph.D. and 60 M.Sc. students as a research advisory member.

Dr. M. Vijaya Lakshmi is working as Technical Officer to Director, PME Cell, Dr.YSRHU. She has 12 years teaching experience and 4 years in research. She has authored 21 research articles, presented 9 research papers in National and International Conferences. She has published a book and one book chapter.

Vallabhapuram Sekhar is working as Assistant Professor in Statistics at Dr. YSRHU - College of Horticulture, Venkataramannagudem, A.P. He has 14 years of teaching experience. He has published 2 text books, 30 research papers in National and International journals.

THE EDITORS

Dr. B. Srinivasulu working as Principal Scientist & Head, Dr.YSRHU-Horticultural Research Station, Ambajipeta and Director of Planning, Monitoring and Evaluation Cell, Dr.YSRHU, A.P with 39 years of experience in Research, Extension activities. He has published 7 text books, 8 book chapters, 147 Research articles, 60 Technical bulletins.

Dr. T. Janakiram working as a Vice-chancellor of Dr. Y. S. R. Horticultural University, A.P. He developed and released 26 different flower crop varieties and 15 hybrid flower crop varieties. He is the recipient of 15 awards at national and international level and fellow of six societies. He has guided 2 Ph.D and 8 M.Sc students. He has outstanding contribution in horticulture research and development in his 32 years of career with impressive record of 260 publications.

STATISTICS FOR HORTICULTURAL SCIENCE

Authors

Dr. K. Umakrishna

Dr. M. Vijaya Lakshmi

Vallabhapuram Sekhar

Edited by

Dr. B. Srinivasulu

Dr. T. Janakiram

2025

Daya Publishing House®

A Division of

Astral International Pvt. Ltd.

New Delhi – 110 002

ISBN: 978-93-5919-359-5

Published by : **Daya Publishing House®**
A Division of
Astral International Pvt. Ltd.
– ISO 9001:2015 Certified Company –
4736/23, Ansari Road, Darya Ganj
New Delhi-110 002
Ph. 011-43549197, 8130496929
E-mail: info@astralint.com
Website: www.astralint.com

DR. Y. S. R. HORTICULTURAL UNIVERSITY

www.drysrhu.ap.gov.in

Dr. T. JANAKIRAM
ARS, PhD (IARI), FISOH,
FISVS, FHSI & FISHRD
Vice-Chancellor

Message

India is currently producing about 320.48 million tonnes of Horticulture produce and has surpassed the food grain production of 315.7 million tonnes. India is world leader in the production of a variety of fruits viz., Mango, Guava, Papaya, Pomegranate, Lime, Aonla and is the second largest producer of fruits and vegetables and maintained its dominance in the production of Coconut, Cashewnut and Spices. In recent days we hear talking about "Statistics" from a common man to highly qualified person. This shows how close "Statistics" is connected with wide range of activities.

This book deals with Statistical tools and techniques for Horticultural science, which includes Collection, Classification, Frequency distribution, Data Representation, Diagrammatic Representation, Measures of Central Tendency, Measures of Dispersion, Elementary Probability, Probability Distributions viz., Binomial Distribution, Poisson Distribution, Normal Distribution, Testing of Hypothesis *viz.*, Concepts, Large sample tests, Small sample tests, Correlation and Regression, Experimental Designs *viz.*, Concepts, Completely Randomized Design, Randomized Block Design and Latin Square Design, with special reference to Horticulture.

This book will be useful for the students and scientists dealing with Horticulture crops and to gain some mastery over these, as well as to learn how to conduct experiments and analyse data. All the usual methodologies are covered in a typical way with practical approach and justification. Besides this book contains various computational techniques in applications.

I Congratulate the authors, Dr. K. Umakrishna, Professor (Statistics), Dr. YSRHU-COH, Venkataramannagudem, Dr. M. Vijaya Lakshmi TO to Director, Planning and Monitoring Cell, Dr. YSRHU and Sri Vallabhapuram Sekhar Assistant Professor (Statistics), Dr. YSRHU-COH, Venkataramannagudem, and the editor, Dr. B. Srinivasulu Director, Planning and Monitoring cell, Dr. YSRHU for excellent drafting and editing information on data collection, calculation and interpretation of statistics on horticultural crops and bringing out as text book entitled "Statistics for Horticultural Science".

(Dr. T. JANAKIRAM)

Administrative Office, Venkataramannagudem – 534101,
West Godavari District, Andhra Pradesh.
Ph: 08818-284312, Fax : 08818-284223, Mobile : 73826 33666,
Email : vc@drysrhu.edu.in

PREFACE

We, the authors believe that, the text book "Statistics for Horticultural Science" is essential for students, research scholars and scientists. This text book contains large number of problems taken from different horticultural crops with complete solutions. The overwhelming response to our text book has prompted us to write and place in the hands of the readers. The contents of this text book has been designed and customized to meet the long standing requirements of those appearing for M.Sc., Ph.D., M.B.A., M. Com., B.Sc., B.Com., ARS, NET, C.A., I.C.W.A., CAT and ICAR Examinations.

Efforts have been made to simplify the technical material without distorting it. The book does not require a high level of mathematical knowledge on the part of the reader. Basic knowledge in fundamental concepts of mathematics is required to understand this text book.

The text book contains ten chapters. The first chapter describes Introduction and Role of statistics in Horticultural science. The second chapter covers collection, classification and tabulation of the horticultural data. Third chapter focuses on graphical and diagrammatic representation of horticultural data. Central tendency measures are covered in fourth chapter and dispersion measures are covered in fifth chapter. Probability and probability distributions are explained in sixth and seventh chapters. The terminology of testing of hypothesis, small sample and large sample tests are demonstrated in eighth chapter. In ninth chapter, correlation and regression concepts analysed. In the tenth chapter, the concepts of designs of experiments and basic designs of experiments problems depicted.

Every effort has been made to present the material in easy, clear, lucid and systematic manner. All efforts have been made to make the book reasonably exhaustive and comprehensive. Valuable suggestions for further improvement of the book will be gratefully accepted.

Authors

CONTENTS

CHAPTER-1
INTRODUCTION

There is a universal acceptance of Statistics as an essential tool for all types of research. In the modern scenario, statistics is very useful subject for students, research scholars, scientists and data analysts. Statistics can be used in plural and singular senses. When it is used as plural form, it is a systematic presentation of facts and figures. These figures may be with regard to production of food grains in different years, area under cereal crops in different years, per capita income in a particular state at different times etc., and these are generally published trade journals, economics and statistics bulletins, news papers etc.. When statistics used as singular, it is a science which deals with collection, classification, tabulation, analysis and interpretation of data. The word Statistics has been derived from the Latin word "Status" which means 'state' which in turn means 'politically organized people'.

R. A. Fisher, father of Statistics conveys that 'Statistics' has three important functions to play (i) Investigation of statistical populations (ii) Investigation of the variation within the statistical populations (iii) Investigation of the methods of reduction of data

P.C. Mahalanobis compares 'Statistician' with a 'Doctor' where Doctor prescribes medicine according to the disease of patient where as statistician suggests suitable statistical technique according to the data in hand for proper analysis and interpretation.

In modern world, the importance of statistics is very well recognized by all the disciplines like Agriculture, Horticulture, Veterinary Science, Medical Science, Engineering, Pharmacy, Management studies and etc.

1.1 The Role of Statistics in Horticultural Research

Statistical methods are now used practically in all branches of horticulture and In fact, it was only due to the need of statistical tools to handle problems in horticultural research, that there was a tremendous development in statistical theory.

Statistics is used in horticultural research for efficient planning of experiments, and for interpreting experimental data.

In horticultural research, efforts are made to know the hidden regularities of some aspects of soils, crops and other related biological phenomena. It may be yield potentials of a crop, effects of pest incidence on crop yields, effects of changes in climate on pests and crops effect of fertilizer application on crop yields, effect of cross-fertilization of different crop varieties, and the like. The present knowledge concerning these aspects may be inadequate. For improving our knowledge about these aspects may take up investigations. Through these investigations an existing theory may be scrutinized or a new theory may be formulated.

At the first stage of a scientific inquiry, data pertaining to the study has to be collected. This process of data collection may involve field trails, laboratory experiments and sample surveys. Statistics comes to the help of the researcher in determining the kind of information, extent of data and method of data collection. There are two branches of statistics, namely, *Experimental Design* and Sampling Design which deal with data collection. In sampling design, the data are recorded without interfering the process that is being observed. In experimental design, the subjects constituting the population are put into controlled environment.

The salient features and significance of the collected data must then be assessed and interpreted through careful analysis. For this purpose, the data may be organized and summarized. The branch of statistics dealing with summarization of data called Descriptive Statistics. The analysis and interpretation of data is dealt with by *Inferential Statistics*. The electronic computers are of great help in processing and analyzing data.

All these branches of Statistics are interlinked. For proper use of statistical techniques, basic data should be valid. A poorly designed and carelessly executed experiment will yield invalid data. In some cases, the collected data may be incomplete or manipulated. The statistical analysis based on such data will lead to illogical or wrong conclusions, no matter how good and sophisticated the statistical tools are. On the other hand, the collected data might be of good quality. But if wrong statistical tools are used to analyze such data, the resulting conclusions might be invalid. The statistical tools may be misused if the user is incompetent. All these aspects require an intensive care. A research worker should be aware of statistical techniques, and he should consult a statistician at all stages of his investigation. The statistician, on the other hand, should be familiar on the experimental subject matter to which statistics is to be applied. This will help in ensuring better coordination in research and improving the quality of research.

Limitations of Statistics

The scope of statistics is very wide and it has great utility, but these are restricted by its limitations. The important statistical limitations are:

- ☆ Statistics does not deal with individual items.
- ☆ Statistics does not deals with qualitative data.
- ☆ Statistical laws are true only on averages
- ☆ Statistics does not reveal the entire theory
- ☆ Statistics is liable to misused.

Application of Statistics in Horticulture

The major applications of statistics in Horticulture are

- ☆ Conventional and molecular breeding
- ☆ Natural Resources
- ☆ Post Harvest Studies
- ☆ Insect and Disease management
- ☆ Climate change studies
- ☆ Social Science Research

CHAPTER-2

Collection, Classification and Frequency Distribution

2.1. COLLECTION OF DATA

Basically collections of data are of two kinds: (i) Primary data (ii) Secondary data.

Primary data is based on primary source of information and the secondary data is based on secondary source of information.

2.1.1. Primary Data is Collected by the Following Methods

(i) Investigator himself conducting survey (ii) Conducting survey with the help of field investigators (iii) Sending questionnaires by post or through mail (iv) Schedules through enumerators

(i) Investigator Himself Conducting Survey:

This method is of much use only in small pilot surveys like case studies. This method is being adopted by individual Investigators who submit dissertation for Master's and Doctoral degrees in Fruit Science, Vegetable Science, Floriculture & Landscape Architecture, Plantation, Spices, Medicinal and Aromatic Crops, Post Harvest Technology, Horticultural Plant Pathology and Horticultural Entomology.

(ii) Conducting Survey with the Help of Field Investigators:

Comprehensive information will be framed by the Chief Investigator with the help of other experts based on objectives of the survey.

(iii) Sending Questionnaires by Post or Through Mail

This method is fast informative for collecting the data with less expenditure but the respondents must be educated.

(iv) Schedules Through Interviewers

The enumerators personally contact the informants, obtain replies to the questions contained in a schedule.

2.1.2. Secondary Data is Collected by the Following Methods:

Existing data collected from news papers, records, third party, State Bureau of Economics and Statistics, Horticulture Statistics at a Glance are some of the main sources of information providing secondary data.

2.2 CLASSIFICATION OF DATA

The major classifications of data are:

(i) Geographical classification

(ii) Chronological classification

(iii) Conditional classification

(iv) Qualitative classification

(v) Quantitative classification.

(i) Geographical Classification:

Here the data is classified according to geographical area like villages, cities, states, countries etc.

For Example : State wise production of potato during the period 2021-22

Name of the State	Production of Potato (tonnes)
Andhra Pradesh	41.27
Madhya Pradesh	3582.00
Uttar Pradesh	15892.00
Kerala	5.36

(ii) Chronological Classification

Here classification is done on the basis of time like hourly, daily, weekly, monthly, quarterly, yearly etc.

For Example : Year wise (2014-2020) Brinjal crop data in Andhra Pradesh

Year	Area	Production
2014	17178.00	378987.00
2015	17061.70	419062.20
2016	15287.00	382175.00
2017	19620.00	510120.00
2018	13707.00	356382.00

2019	15505.00	403130.00
2020	15027.82	390728.30

(iii) Conditional Classification

Here data is classified on certain conditions.

Example : Student performance in the subject Statistics

Grade point(G.P.) in Statistics	Result
< 5.0	Fail
5.0-8.0	First Class
>8.0.	Distinction

(iv) Qualitative Classification

Here data is classified on the basis of some attributes like color, shapes etc.

Example : Leaf shape

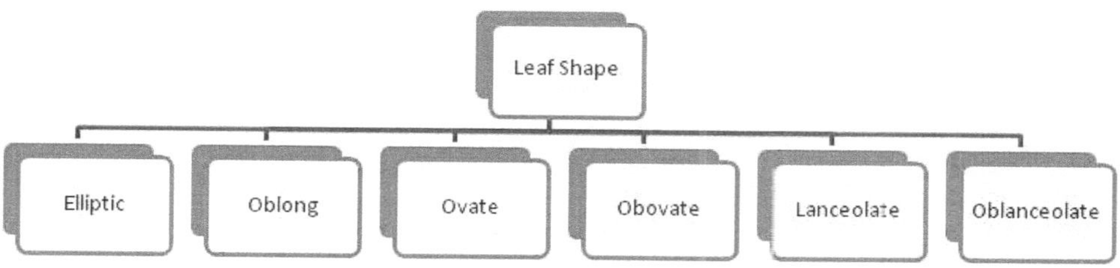

(v) Quantitative Classification

When the data is classified on the basis of the numerical measurements is called as Quantitative classification. It can be measured as height, weight, production, sales etc.

Example : If we take 100 mango fruits and measured its weight

Weight	No. of mangoes
0-20	40
20-40	30
40-60	20
60-80	6
80-100	4

2.3 FREQUENCY DISTRIBUTION

A particular value is occurred in number of times is called as frequency. It is simply indicated by 'f'. Set of real values of random variable and their frequencies is called frequency distribution.

Example:

If the plant height lies between 6ft to 8ft, the frequency for the 6-8 intervals is 2.

Frequency distribution has two types

1. Discrete Frequency distribution 2. Continuous Frequency distribution

2.3.1 Discrete Frequency Distribution

The distribution of frequencies according to discrete variable is called discrete frequency distribution. If the variable takes a countable number of values, then it is easier to present and handle the data using by discrete frequency distribution.

Example:

In each of 15 plants marked rugose spiraling whitefly in different places in west Godavari district. The results were recorded as follows:

5,4,2,4,3,5,0,1,5,4,3,0,1,2,4

Discrete frequency table for the number of plants marked rugose spiraling whitefly in different places in west Godavari district

Rugose spiraling whitefly (X)	0	1	2	3	4	5
Number of plants marked as rugose spiraling whitefly (f)	2	2	2	3	4	2

2.3.2 Continuous Frequency Distribution

A variable which assumes the values between two fixed limits is called continuous variable. Distribution of frequencies according to continuous variable is called continuous distribution.

Example:

The following details belongs to students and their corresponding grade points (G.P) in final examination

G.P (X)	0-2	2-4	4-6	6-8	8-10
Number of students	20	15	40	20	5

CHAPTER-3

Data Representation

3.1 GRAPHICAL REPRESENTATION

1. Histogram

When the data is classified based on the class intervals, it can be represented by a histogram. Histogram is just like a simple bar diagram with minor differences. There is no gap between the bars, since the classes are continuous. The bars are drawn only in outline without coloring or marking as in the case of simple bar diagrams. It is the suitable form to represent a frequency distribution. Class intervals are to be presented in x axis and the bases of the bars are the respective class intervals. Frequencies are to be represented in y axis. The heights of the bars are equal to the corresponding frequencies.

Example:

Draw a histogram for the following data:

Seed Yield (gms)	No. of Plants
0.5-2.5	6
2.5-4.5	8
4.5-6.5	15
6.5-8.5	29
8.5-10.5	14
10.5-12.5	7

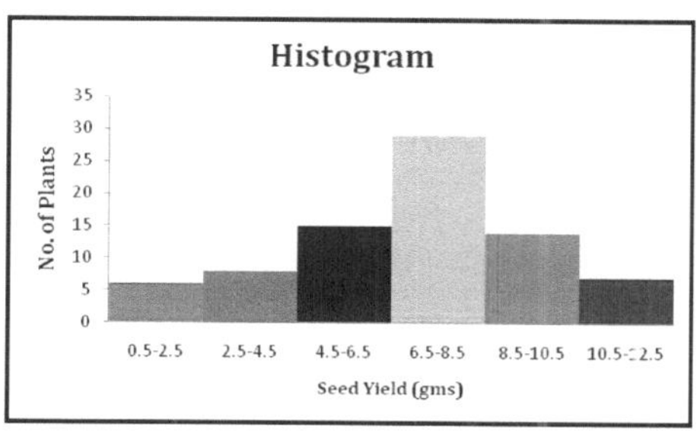

2. Frequency Polygon

The frequencies of the classes are plotted by dots against the mid-points of each class. The adjacent dots are then joined by straight lines. The resulting graph is known as frequency polygon.

Example:

Draw frequency polygon for the following data

Seed Yield (gms)	No. of Plants
0.5-2.5	6
2.5-4.5	8
4.5-6.5	15
6.5-8.5	29
8.5-10.5	14
10.5-12.5	7

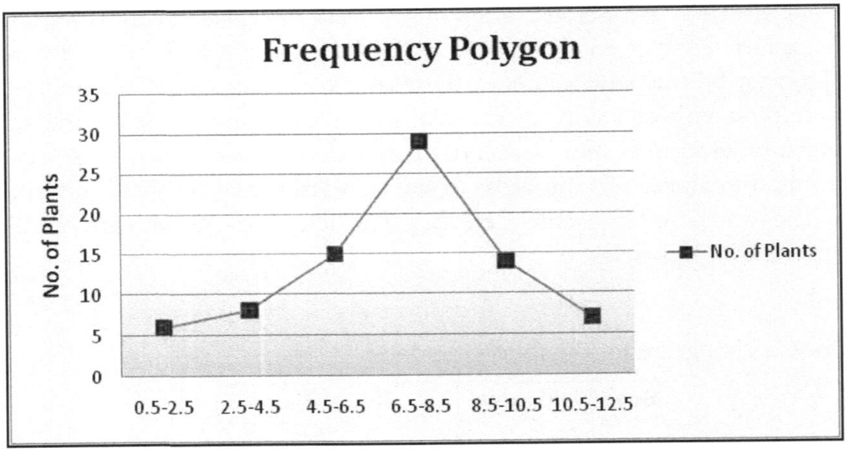

3. Frequency Curve

The procedure for drawing a frequency curve is same as for frequency polygon. But the points are joined by smooth or free hand curve.

Example:

Draw frequency curve for the following data

Seed Yield (gms)	No. of Plants
0.5-2.5	6
2.5-4.5	8
4.5-6.5	15

Seed Yield (gms)	No. of Plants
6.5-8.5	29
8.5-10.5	14
10.5-12.5	7

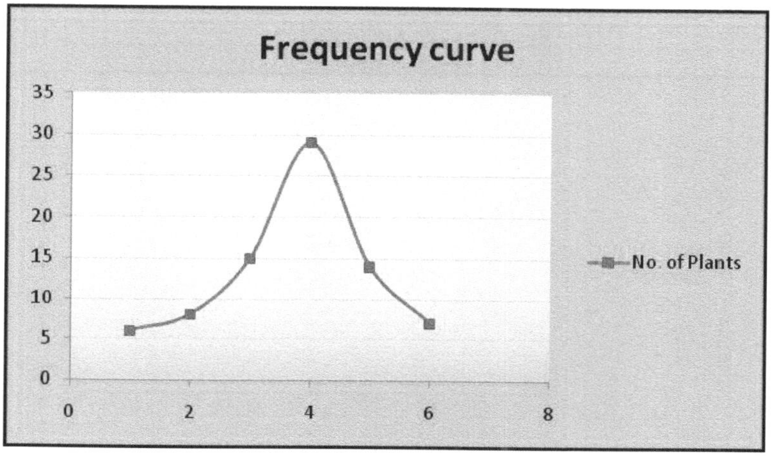

Ogives

Ogives are known also as cumulative frequency curves and there are two kinds of ogives. One is less than ogive and the other is more than ogive.

Less than ogive: Here the less then cumulative frequencies are plotted against the upper boundary of respective class interval.

Greater than ogive: Here the greater than cumulative frequencies are plotted against the lower boundaries of respective class intervals.

Note: X-coordinate of the point of intersection of both cumulative frequency curves is called as Median

Example:

Find the ogive curve for the following seed weight of number of plants

Seed Yield (gms)	0.5-2.5	2.5-4.5	4.5-6.5	6.5-8.5	8.5-10.5	10.5-12.5
No. of Plants	6	8	15	29	14	7

Solution:

Seed Yield (gms)	midpoint	No. of Plants	< cumulative frequency	> cumulative frequency
0.5-2.5	1.5	6	6	79
2.5-4.5	3.5	8	14	73

Seed Yield (gms)	midpoint	No. of Plants	< cumulative frequency	> cumulative frequency
4.5-6.5	5.5	15	29	65
6.5-8.5	7.5	29	58	50
8.5-10.5	9.5	14	72	21
10.5-12.5	11.5	7	79	7

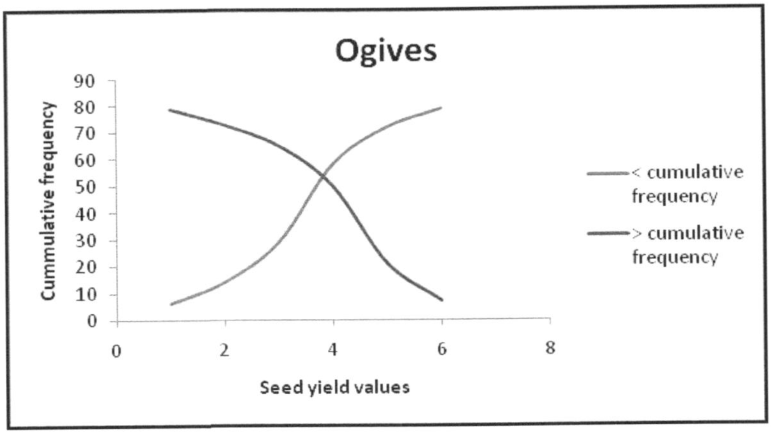

3.2 DIAGRAMMATIC REPRESENTATION

Diagrams

Diagrams are various geometrical shape such as bars, circles etc. Diagrams are based on scale but they are not confined to points or lines. They are more attractive and easier to understand than graphs.

Merits

- ☆ Most of the people are attracted by diagrams.
- ☆ Technical Knowledge or education is not necessary.
- ☆ Time and effort requirements are less.
- ☆ Diagrams show the data in proper perspective.
- ☆ Language is not a barrier.
- ☆ Widely used tool.

Demerits (or) limitations

- ☆ Diagrams are approximations.
- ☆ Minute differences in values cannot be represented properly in diagrams.
- ☆ Large differences in values spoil the appearance of the diagram.

Types of Diagrams

The important diagrams are:

- ☆ Simple Bar diagram.
- ☆ Multiple Bar diagram.
- ☆ Component Bar diagram.
- ☆ Percentage Bar diagram.
- ☆ Pie chart
- ☆ Pictogram
- ☆ Statistical maps or cartograms.

In all the diagrams, the groups or classes are represented on the x-axis and the volumes or frequencies are represented in the y-axis.

1. Simple Bar Diagram

If the classification is based on attributes and if the attributes are to be compared with respect to a single character we use simple bar diagram.

Example

- ☆ The area under different crops in a state.
- ☆ The food grain production of different years.
- ☆ The yield performance of different varieties of a crop.
- ☆ The effect of different treatments etc.

Simple bar diagrams Consists of vertical bars of equal width. The heights of these bars are proportional to the volume or magnitude of the attribute. All bars stand on the same baseline. The bars are separated from each others by equal intervals. The bars may be coloured or marked.

Example:

District wise Horticultural crops area in 1000 acres in Andhra Pradesh for the year 2022-23 was as follows

District Name	Area
Alluri Sitharama Raju	259
Anakapalli	832
Ananthapuramu	13,698
Annamayya	10,754
Bapatla	1,209
Chittoor	6,681
East Godavari	2,941
Eluru	4,658

District Name	Area
Guntur	560
Kakinada	1,053
Konaseema	44
Krishna	280
Kurnool	4,509
Nandyal	4,438
Ntr	1,803
Palnadu	1,582
Parvathipuram Manyam	552
Prakasam	8,851
S.P.S Nellore	3,437
Sri Sathya Sai	9,348
Srikakulam	539
Tirupati	1,632
Visakhapatnam	68
Vizianagaram	891
West Godavari	162
Y.S.R Kadapa	9,982

The simple bar diagram for this data is given below.

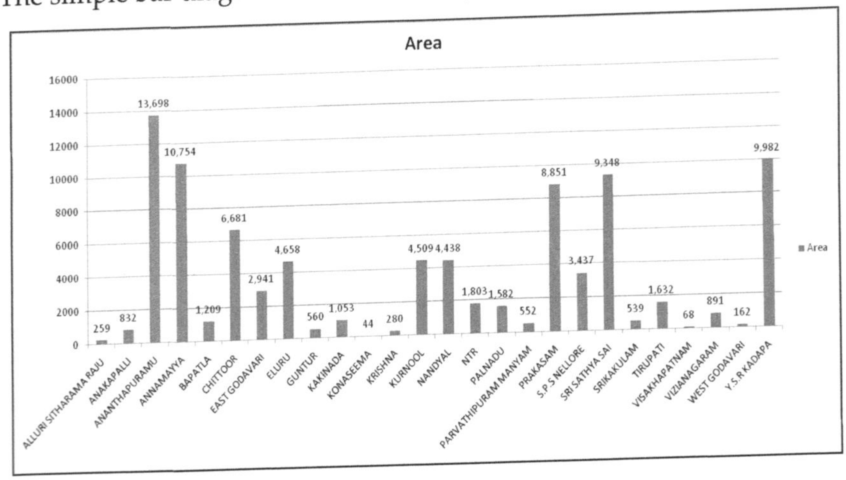

2. Multiple Bar Diagram

If the data is classified by attributes and if two or more characters or groups are to be compared within each attribute we use multiple bar diagrams. If only two characters are to be compared within each attribute, then the resultant bar diagram used is known as double bar diagram. The multiple bar diagram is simply the extension of simple bar diagram. For each attribute two or more bars representing separate characters or groups are to be placed side by side. The Multiple bar diagram for this data is given below.

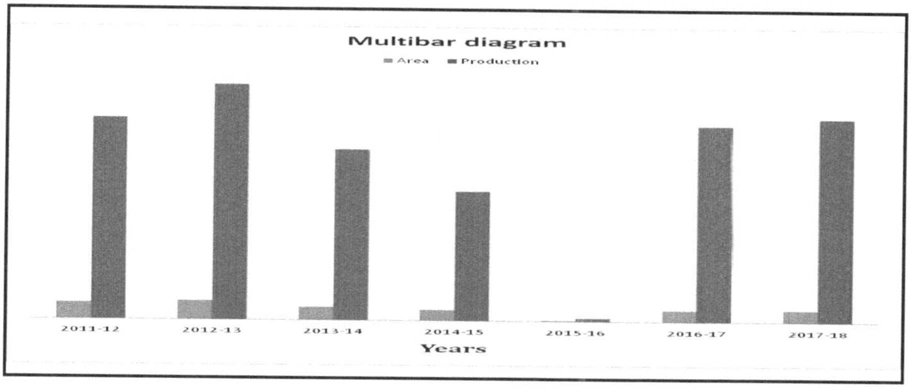

3. Component Bar Diagram

This is also called sub – divided bar diagram. Instead of placing the bars for each component side by side we may place these one on top of the other. This will result in a component bar diagram is given below.

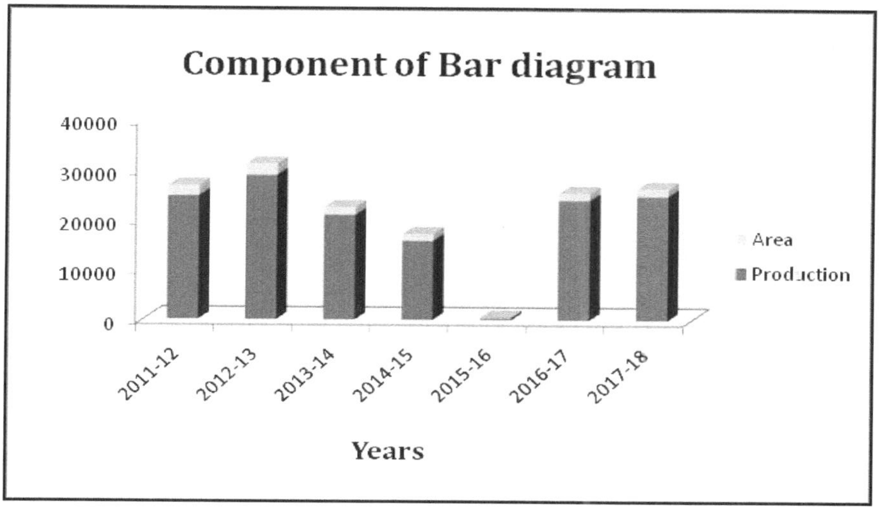

4. Percentage Bar Diagram

Sometimes when the volumes of different attributes may be greatly different for making meaningful comparisons, the attributes are reduced to percentages. In that case each attribute will have 100 as its maximum volume. This sort of component bar chart is known as percentage bar diagram.

$$\text{Percentage} = \frac{\text{Actual value}}{\text{Total of the actual value}} \times 100$$

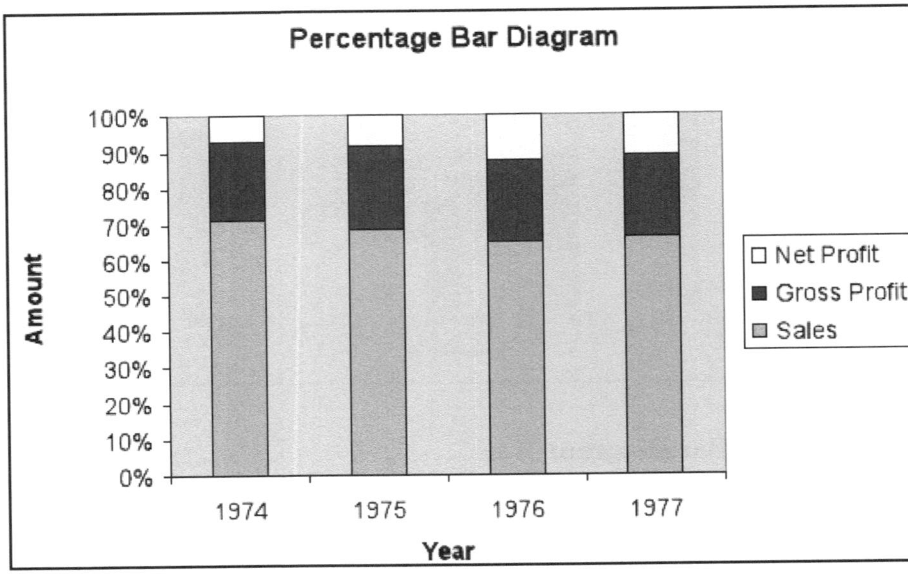

5. Pie Chart / Pie Diagram

Pie diagram is a circular diagram. It may be used in place of bar diagrams. It consists of one or more circles which are divided into a number of sectors. In the construction of pie diagram the following steps are involved.

Step 1:

Whenever one set of actual value or percentage are given, find the corresponding angles in degrees using the following formula

$$\text{Angle} = \frac{\text{Actual value}}{\text{Total of the actual value}} \times 360^0$$

Step 2:

Find the radius using the area of the circle πr^2 where value of π is $\frac{22}{7}$ or 3.14

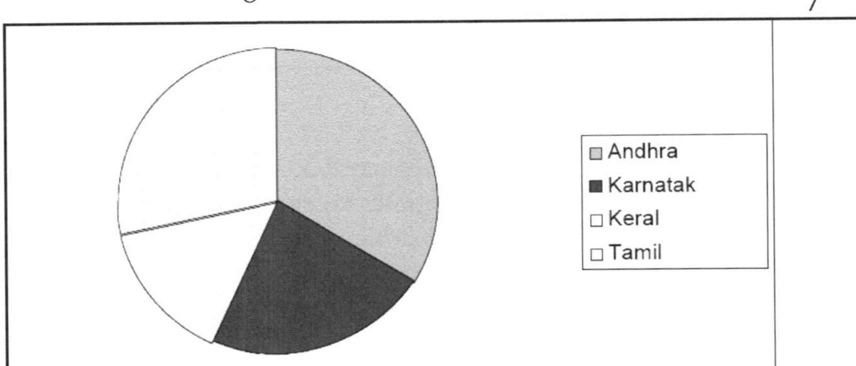

CHAPTER -4

Measures of Central Tendency (Averages)

Definition

A Single value which represents group of values is called as central tendency. It has five types

1. Arithmetic Mean
2. Median
3. Mode
4. Geometric Mean
5. Harmonic Mean

4.1 ARITHMETIC MEAN

Arithmetic mean of a variable is defined as the sum of the observations divided by the number of observations. It is denoted by the symbol $\overline{x}$. If its variable assumes n values $x_1, x_2 \ldots x_n$ then the mean is given by

Ungrouped or Raw Data

$$\overline{x} = \frac{X_1 + X_2 + \ldots + X_n}{n}$$

$$\frac{1}{n}\sum_{i-1}^{n} x_i$$

n→ Number of observations

Example:

Calculate the mean for pH levels of soil 6.8, 6.6, 5.2, 5.6, 5.8

$$\overline{X} = \frac{6.8+6.6+5.2+5.6+5.8}{5} = \frac{30}{5} = 6$$

Grouped Data

The mean for grouped data is obtained from the following formula:

$$\overline{X} = \frac{f_1x_1 + f_2x_2 + \ldots + f_nx_n}{N}$$

$$\overline{X} = \frac{\sum_{i=1}^{n} f_i x_i}{N}$$

Where X = Mid value of Class Interval

N = Total frequency

Deviation Method (or) Linear Transformation Method or Short-Cut Method

$$\overline{X} = A + \frac{\sum_{i=1}^{n} f_i d_i}{N} \times h$$

A = any value in x

N = Total frequency

h = Width of the class interval

$$d_i = \frac{X_i - A}{h}$$

Merits

- ☆ It is rigidly defined.
- ☆ It is easy to understand and easy to calculate.
- ☆ If the number of items is sufficiently large, it is more accurate and more reliable.
- ☆ It is a calculated value and is not based on its position in the series.
- ☆ It is possible to calculate even if some of the details of the data are missing.
- ☆ It is affected least by fluctuations of sampling.
- ☆ It provides a good basis for comparison.

Demerits

- ☆ It cannot be obtained by inspection nor located through a frequency graph.
- ☆ It cannot be in the study of qualitative phenomena and not capable of numerical measurement i.e. Intelligence, beauty, honesty etc.

- ☆ It can ignore any single item only at the risk of losing its accuracy.
- ☆ It is affected very much by extreme values.
- ☆ It cannot be calculated for open-end classes.
- ☆ It may lead to fallacious conclusions, if the details of the data from which it is computed are not given.

4.2 MEDIAN

The median is the middle most value that divides the group into two equal parts, one part comprising all values greater, and the other, all values less than that value.

Ungrouped or Raw Data

Step1 : Arrange the given values in the ascending order.

Step2 : If the number of values are odd, median is the middle value. If the number of values are even, median is the mean of middle two values.

By formula,

When n is odd, Median = Md = $\left(\frac{n+1}{2}\right)^{th}$ value

When n is even, Average of $\left(\frac{n}{2}\right)^{th}$ and $\left(\frac{n}{2}+1\right)^{th}$ value

Example:

If the weights of tomato are 45, 60, 48, 100, 65 in gms, calculate the Median

Solution:

Here n = 5,

First we arrange it in ascending order 45, 48, 60, 65, 100

$$\text{Median} = \text{Md} = \left(\frac{n+1}{2}\right)^{th} \text{ value}$$

$$= \left(\frac{5+1}{2}\right)^{th} = 3^{rd}\,value = 60\,g$$

Example:

If the weights of tomato are 45, 48, 60, 65, 65, 100 in gms, calculate the median.

SolutionHere n = 6

Median = Md =Average of $\left(\frac{n}{2}\right)^{th}$ and $\left(\frac{n}{2}+1\right)^{th}$ values = Average of $\left(\frac{6}{2}\right)^{th}$ and $\left(\frac{6}{2}+1\right)^{th}$ value

=Average of 3rd and 4th values $= \frac{60+65}{2} = 62.5$ g

Grouped Data

In a grouped distribution, values are associated with frequencies. Grouping can be in the form of a discrete frequency distribution or a continuous frequency distribution. Whatever may be the type of distribution, cumulative frequencies have to be calculated to know the total frequency.

Cumulative Frequency (CF)

Cumulative frequency of each class is the sum of the frequency of the class and the frequencies of the pervious classes, i.e. adding the frequencies successively, so that the last cumulative frequency gives the total frequency.

Step 1 : Find cumulative frequencies (cf).

Step 2 : Find $\frac{N}{2}$; N = Total frequency.

Step 3 : See in the cumulative frequency the value first greater than N/2, Then the corresponding class interval is called the Median class. Then apply the formula

$$\text{Median} = l + \frac{\frac{N}{2} - cf}{f} X h$$

Where

l = Lower limit of the median class

cf = cumulative frequency preceding the median class

h = width of the class

f = frequency in the median class.

N = Total frequency.

Merits of Median

- ☆ Median is not influenced by extreme values because it is a positional average.
- ☆ Median can be calculated in case of distribution with open-end intervals.
- ☆ Median can be located even if the data are incomplete.

Demerits of Median

- ☆ A slight change in the series may bring drastic change in median value.
- ☆ In case of even number of observations or continuous series, median is an estimated value other than any value in the series.
- ☆ It is not suitable for further mathematical treatment except its use in calculating mean deviation.
- ☆ It does not take into account all the observations.

4.3 MODE

- ☆ The mode refers to that value in a distribution, which occur most frequently. It is an actual value, which has the highest concentration of items in and around it.
- ☆ It shows the centre of concentration of the frequency in around a given value. Therefore, where the purpose is to know the point of the highest concentration it is preferred. It is, thus, a positional measure.
- ☆ Its importance is very great in agriculture like to find typical height of a crop variety, maximum source of irrigation in a region, maximum disease prone paddy variety. Thus the mode is an important measure in case of qualitative data.

Computation of the Mode

Ungrouped or Raw Data

For ungrouped data or a series of individual observations, mode is often found by mere inspection.

Find the mode for the following seed weight 2 , 7, 10, 15, 10, 17, 8, 10, 2 gms

$\therefore$ Mode = 10

In some cases the mode may be absent while in some cases there may be more than one mode.

Example:

(1) 12, 10, 15, 24, 30 (no mode)

(2) 7, 10, 15, 12, 7, 14, 24, 10, 7, 20, 10

The modal values are 7 and 10 as both occur 3 times each.

Grouped Data

Locate the highest frequency the class corresponding to that frequency is called the modal class. Then apply the formula.

$$\text{Mode} = l + \frac{f_1 - f_0}{(f_1 - f_0) - (f_2 - f_1)} \times h$$

Where

l = lower limit of the model class

f_0 = the frequency of the class preceding the modal class

f_1 = the frequency of the modal class

f_2 = the frequency of the class succeeding the modal class and

h = width of class interval

Applications

- ☆ It is easy for interpretation and calculation. The mode is mostly used in business forecasting.
- ☆ Mean calculation is always done by all observations. So the mode is not affected by extreme values or abnormal data.

Merits

- ☆ It can be easily located on frequency curve
- ☆ It can be calculated even when extreme classes are not well defined except the modal class
- ☆ It is used mostly in business For Example , Cloth merchant would like to keep a certain quality of cloth having maximum sales, shoe-maker would like to keep a shoe of size or sizes having maximum sales, and
- ☆ It can easily be calculated except in the case where the maximum frequency occurs more than once.

Demerits

- ☆ It is not based upon all the observations
- ☆ It is not having algebraic properties, and
- ☆ It is not stable, since different methods of forming class intervals would lead to different modal values.

Median and Mode for Grouped Data

The following is the distribution of heights of 100 Tiger lily plants

Height(cms.)	10-20	20-30	30-40	40-50	50-60	60-70
No.of Plants	10	15	28	30	10	7

Solution:

MEDIAN

Aim: Calculate the Median and Mode for heights of 100 Tiger lily plants Median and Mode

Formula and Procedure:

Cumulative frequency of each class is the sum of the frequency of the class and the frequencies of the previous classes, i.e. adding the frequencies successively, so that the last cumulative frequency gives the total frequency.

Step 1 : Find cumulative frequencies.

Step 2 : Find $\frac{N}{2}$; N = Total frequency.

Step 3 : See in the cumulative frequency the value first greater than, Then the corresponding class interval is called the Median class. Then apply the formula

$$\text{Median} = 1 + \frac{\frac{N}{2} - \text{cf}}{\text{f}} \text{XH}$$

Where

l = Lower limit of the medianal class

cf = cumulative frequency preceding the median class

h = width of the class

f =frequency in the median class.

N=Total frequency.

Calculation:

Height (cms.)	No.of Plants	Cummulative Frequency(c.f.) Lessthan
10-20	10	10
20-30	15	25(cf)
(L)30-40	28(f)	53
40-50	30	83
50-60	10	93
60-70	7	100

N/2 = 100/2 = 50

$$\text{Median} = 1 + \frac{\frac{N}{2} - \text{cf}}{\text{f}} \times \text{h} = 33 + \frac{50-25}{28} \times 10 = 41.9286$$

MODE:

Formula and Procedure:

Locate the highest frequency the class corresponding to that frequency is called the modal class. Then apply the formula.

$$\text{Mode} = l + \frac{f_1 - f_0}{(f_1 - f_0) - (f_0 - f_2)} \times h$$

Where

l = Lower limit of the modal class

f_0 = The frequency of the class preceding the modal class

f_1 = The frequency of the modal class

f_2 = The frequency of the class succeeding the modal class and

h = Length of class interval

Calculation:

Height(cms.)	No.of Plants
10-20	10
20-30	15
30-40	28(f0)
(L)40-50	30(f1)
50-60	10(f2)
60-70	7

$$\text{Mode} = l + \frac{f_1 - f_0}{(f_1 - f_0) - (f_0 - f_2)} \times h$$

$$= 40 + \frac{2}{2-(-20)} \times 10 = 40 + \frac{20}{22} = 40.9091$$

Inference:

The Median height of 100 Tiger lily plants = 41.9286

The modal height of 100 Tiger lily plants = 40.9091

Note:

The formula for empirical relationship between mean, median and mode is

Mode = 3Median - 2Mean

4.4 GEOMETRIC MEAN

The geometric mean of a series containing n observations is the nth root of the product of their values. If $x_1, x_2, \ldots, x_n$ are observations then

For Ungrouped Data

$$\text{G. M.} = \sqrt[n]{x_1 . x_2 \ldots x_n} \quad \ldots (1)$$

$$= (x_1 . x_2 \ldots x_n)^{\frac{1}{n}} \quad \ldots (2)$$

Take the "log" on both side for equation (2), we get

$$\text{Log G.M.} = \frac{1}{n} \log(x_1 . x_2 \ldots x_n)$$

$$= \frac{\sum_{i=1}^{n} log\, x_i}{n}$$

$$\text{G.M.} = \text{Antilog}\left(\frac{\sum_{i=1}^{n} log\, x_i}{n}\right) \quad \ldots (3)$$

For Grouped Data

$$\text{G.M.} = \text{Antilog}\left(\frac{\sum_{i=1}^{n} f_i . \log x_i}{N}\right) \qquad \ldots (4)$$

Example:

If the nutrition quantity in five mushrooms are 45, 60, 48,100, 65 in gms. Then find the Geometric mean?

Solution:

Nutrition quantity in mushrooms (x_i)	Log (x_i)
45	1.653
60	1.778
48	1.681
100	2
65	1.8153
Total	8.925

Here n = 5

$$\text{G.M.} = \text{Antilog}\ \frac{\sum_{i=1}^{n} \log x_i}{n}$$

$$= \text{Antilog}\ \frac{8.925}{5}$$

$$= \text{Antilog}\ 1.785$$

$$= 60.95$$

Grouped Data

Example:

Find the Geometric mean for the following data

Nutrition quantity in mushrooms (x_i)	Frequency (f_i)
50	4
65	6
75	16
80	8
95	7
100	4

Solution:

Nutrition quantity in mushrooms (x_i)	Frequency (f_i)	Log	f_i log (x_i)
50	5	1.699	8.495
63	10	10.799	17.99
65	5	1.813	9.065
130	15	2.114	31.71
135	15	2.13	31.95
Total	**50**	**9.555**	**99.21**

Here N = 50 = total frequency.

$$\text{G.M.} = \text{Antilog } \frac{\sum_{i=1}^{n} f_i .\log x_i}{N} = \text{Antilog } \frac{99.21}{50} = \text{Antilog } 1.9842 = 96.43$$

Application of G.M.

GM is used in studies like growth rate of chronological yield data, bacterial growth, cell division, etc.

Merits

- ☆ It has we1l defined.
- ☆ It is based upon all the observations.
- ☆ It is used in computing index numbers and also in time series analysis whenever ratios are under consideration. It is also used in finding out rate of change in population and computing compound interest.
- ☆ It possesses algebraic properties.

Demerits

- ☆ It is difficult to calculate.
- ☆ It cannot be calculated whenever zero value is present in the observations.
- ☆ It may not be identified with any of the given observations.

4.5 HARMONIC MEAN (H.M.)

Harmonic mean of a set of observations is defined as the reciprocal of the arithmetic average of the reciprocal of the given values. If $x_1, x_2.....x_n$ are n observations,

$$\text{H.M.} = \frac{n}{\sum_{i=1}^{n}\left(\frac{1}{x_i}\right)} \quad (1)$$

For a frequency distribution

$$\text{H.M.} = \frac{N}{\sum_{i=1}^{n} f_i\left(\frac{1}{x_i}\right)} \quad (2)$$

Application of H.M.:

H.M. is used when we are dealing with speed, rates, etc.

Example

For Ungrouped data:

From the given data 5, 10, 17, 24, 30 calculate H.M.

x	1/x
5	0.2
10	0.1
17	0.05
24	0.04
30	0.4

$$\text{H.M.} = \frac{n}{\sum_{i=1}^{n}\left(\frac{1}{x_i}\right)}$$

$$\text{H.M.} = \frac{5}{0.4338} = 11.526$$

Example:

Grouped Data

Find the Harmonic mean for the following

Nutrition quantity in mushrooms (x_i)	Frequency (f_i)
50	4
65	6
75	16
80	8
95	7
100	4

Solution:

Nutrition quantity in mushrooms (x_i)	Frequency (f_i)	$1/x_i$	f_i*($1/x_i$)
50	4	0.02	0.08
65	6	0.015	0.09
75	16	0.01	0.213
80	8	0.0125	0.1
95	7	0.0105	0.074
100	4	0.01	0.04

$$\text{H.M.} = \frac{N}{\sum_{i=1}^{n} f_i\left(\frac{1}{x_i}\right)} = \frac{45}{0.597} = 73.376$$

Merits

- ☆ It is rigidly defined.
- ☆ It is defined on all observations.
- ☆ It is amenable to further algebraic treatment.
- ☆ It is the most suitable average when it is desired to give greater weight to smaller observations and less weight to the larger ones.

Demerits

- ☆ It is not easily understood.
- ☆ It is difficult to compute.
- ☆ It is only a summary figure and may not be the actual item in the series
- ☆ It gives greater importance to small items and is therefore, useful only when small items have to be given greater weightage.
- ☆ It is rarely used in grouped data.

CHAPTER-5

MEASURES OF DISPERSION

INTRODUCTION

- ☆ Dispersion is the measure to extend to which individual items vary.
- ☆ It is our common experience that there are certain varieties of a particular crop, which are very responsive to doses of inputs, and others are not.
- ☆ If these input responsive varieties are provided with a high dose of nutrient they come up with very good yields, on the other hand, if these varieties are put under input stressed condition then their performance will be very poor.
- ☆ Like the measures of Central Tendency, there is a need to have measures for dispersion also.
- ☆ In fact, different measures of dispersions are available in the theory of statistics. Before going details into the discussion of different measures of dispersion, let us try to examine the characteristics of a good measure of dispersion. A good measure of dispersion should have the following characteristics:
  - ☆ It should be defined clearly.
  - ☆ It should be convincing and easy to understand.
  - ☆ It should be easy for calculation and mathematical treatment.
  - ☆ It should be based on all observations.
  - ☆ Good evaluate of dispersion should be least affected by the sampling fluctuations.
  - ☆ It should not be affected badly by the extreme observation.
  - ☆ Generally, there are 6 types of measure of Dispersion

Range

1. Quartile Deviation
2. Mean Deviation
3. Standard Deviation
4. Variance
5. Coefficient of variance

5.1 RANGE

It is defined as the difference between the highest and lowest values in a series of observations.

Example:

Find the 'Range' for the following weights of 15 tomatos in grams is as follows:

50, 65, 75, 45, 55, 45, 60, 110, 95 30, 80, 85, 50, 45, 60.

Solution:

Range: 110-30=80.

Merits

- ☆ This measure is easy to compute.
- ☆ This is useful when the data are of homogeneous nature.
- ☆ It is also used in the preparation of control charts and for the data based on daily temperatures, rainfall, etc.

Demerits

- ☆ Range is not much used in practice since it depends upon two extreme values.
- ☆ Presence of any extremely high and low values in the observations will affect the range considerably.

5.2 QUARTILE DEVIATION: (SEMI-INTER QUARTILE RANGE)

The quartiles divide the distribution in four parts. There are three quartiles. The second quartile divides the distribution into two halves and therefore is the same as the median. The first quartile (Q1), the third quartile (Q3). It may be noted that the second quartile is the value of the median.

It is given by the formula,

$$\text{Q.D.} = \frac{(Q_3 - Q_1)}{2}$$

where Q_1 = First quartile,

Q_3 = Third quartile.

☆ First and third quartiles are also called as lower and upper quartiles respectively.

CASE 1

RAW (or) UNGROUPED DATA

Step 1 : Arrange the given data into the increasing or decreasing order

Step 2 : Use the formula for Q_1 and Q_3

Step 3 : Quartile Deviation,

$$Q.D. = \frac{(Q_3 - Q_1)}{2}$$

$$Q_1 = \frac{n+1}{4} \quad Q_3 = \frac{3(n+1)}{4}$$

Ungrouped Data

Example :

Find the Q.D. for the following observations on number of mesta plants in 10 equi-sized plots. 13, 9, 16, 4, 8, 19, 7, 23, 21, 12.

Step 1 : Arranging the values in ascending order of magnitude, we have 4, 7, 8,9, 12, 13, 16, 19, 21, 23.

Step 2 : $Q_1 = \frac{n+1}{4} = \frac{10+1}{4} = 2.75$

Q_1 = 2nd value+0.75 (3rd value-2nd value)

= 7+0.75x 1=7.75

$$Q_3 = \frac{3(n+1)}{4} = \frac{33}{4} = 8.25$$

Q_3 = 8th value + 0.25 (9th value - 8th value)

= 19+0.25 (21-19)= 16.25

$$Q.D. = \frac{16.25-7.75}{2} = 4.25$$

CASE (ii)

Continuous Distribution

Step 1 : Find cumulative frequencies

Step 2 : Find $\frac{N}{4}$

Step 3 : See in the cumulative frequencies, the value just greater than $\frac{N}{4}$, then the corresponding class interval is called first quartile class.

Step 4 : Find See in the cumulative frequencies the value just greater than then the corresponding class interval is called 3rd quartile class. Then apply the respective formulae.

$$Q_1 = l_1 + \frac{\frac{N}{4} - cf_1}{f_1} * h_1 \quad \text{... (1)}$$

$$Q_3 = l_3 + \frac{\frac{3N}{4} - cf_3}{f_3} * h_3 \quad \text{... (2)}$$

Where l_1 = lower limit of the first quartile class

f_1 = frequency of the first quartile class

h_1 = width of the first quartile class

cf_1 = c.f. preceding the first quartile class

l_3 = lower limit of the third quartile class

f_3 = frequency of the third quartile class

h_3 = width of the third quartile class

cf_3 = c.f. preceding the third quartile class

Merits

- ☆ It can be easily calculated and simply understood.
- ☆ It does not involve much mathematical difficulties.
- ☆ As it takes middle 50% terms hence it is a measure better than Range and Percentile Range.
- ☆ It is not affected by extreme terms as 25% of upper and 25% of lower terms are left out. In case we are to deal with the center half of a series this is the best measure to use.

Demerits

- ☆ Q_1 and Q_3 are both positional measures hence are not capable of further algebraic treatment.
- ☆ Calculations are much more, but the result obtained is not of much importance.
- ☆ It is too much affected by fluctuations of samples.

5.3 MEAN DEVIATION:

It is the mean of the absolute values of the deviations taken from some average or from any point.

Mean Deviation about Mean: Let $X_1, X_2, \ldots, X_n$ be n observations, the Mean deviation about mean or median (md) or mode (m_o) is given by the formula

M.D. about mean = $\frac{\sum_{i=1}^{n}|x_i - \overline{x}|}{n}$

M.D. about Median = $\frac{\sum_{i=1}^{n}|x_i - \text{md}|}{n}$

M.D. about Mode = $\frac{\sum_{i=1}^{n}|x_i - \text{m}_0|}{n}$

n = Number of observations

Example:

1. Find the M.D. about mean for the following data.

4, 9, 10, 14, 7, 8, 6, 14.

Solution:

X_i	4	9	10	14	7	8	6	14	Total
$\lvert X_i - \overline{X}\rvert$	5	0	1	5	2	1	3	5	**22**

$$\overline{x} = \frac{\sum_{i=1}^{n} x_i}{n} = \frac{4+9+10+14+7+8+6+14}{8} = 9$$

M.D. about mean $\frac{\sum_{i=1}^{n}|x_i - \overline{x}|}{n} = 22/8 = 2.75$

2. Find the M.D. about Median for the following data.

4, 9, 10, 14, 7, 8, 6, 14.

Solution:

X_i	4	9	10	14	7	8	6	14	Total
$\lvert x_i - Median\rvert$	6.5	1.5	0.5	3.5	3.5	2.5	4.5	3.5	**26**

Median= (14+7)/2 = 10.5

M.D. about Median = = 26/8 = 3.25

Note:

Mean Deviation about mean is powerful than compare to Mean Deviation about median or Mean Deviation about mode.

Continuous Frequency Distribution

Let f_1, f_2, ... ,fn be n frequencies corresponding to the mid values of the classes X_1, X_2,..,Xn respectively then the M.D. about mean is given by the formula

$$\text{M.D. about mean} = \frac{\sum_{i=1}^{n} f_i |x_i - \overline{x}|}{N} \quad \ldots (4)$$

$$\overline{x} = \frac{\sum_{i=1}^{n} f_i x_i}{N} \quad \ldots (5)$$

N =Total Frequency.

Q. Find the M.D. About Mean to the Following Data

Fruits (in kgs)	C.I	0-2	02-04	04-06	06-08	08-10	10-12	12-14
No. of trees	(f_i)	6	10	14	18	11	7	5

Solution:

Fruits (in kgs)	No. of trees	Mid value	$f_i x_i$	$\|x_i - \overline{x}\|$	$f_i \|x_i - \overline{x}\|$
C.I.	f_i	x_i			
0-2	6	1	6	5.66	33.96
2-4	10	3	30	3.66	36.6
4-6	14	5	70	1.66	23.24
6-8	18	7	126	0.34	6.12
8-10	11	9	99	2.34	25.74
10-12	7	11	77	4.34	30.38
12-14	5	13	65	6.34	31.7
Total	**71**		**473**		**187.74**

$$\overline{x} = \frac{\sum_{i=1}^{n} f_i x_i}{N} = \frac{473}{71} = 6.66$$

N =Total frequency = 71

$$\text{M.D. about mean} = \frac{\sum_{i=1}^{n} f_i |x_i - \overline{x}|}{N} = \frac{187.74}{71} = 2.64$$

Merits or Uses

- ☆ It is simple to understand and easy to compute.
- ☆ It is based on each and every item of the data.
- ☆ MD is less affected by the values of extreme items than the Standard deviation.

Demerits or Limitations or Drawbacks

- ☆ The greatest drawback of this method is that algebraic signs are ignored while taking the deviations of the items.
- ☆ It is not capable of further algebraic treatments.
- ☆ It is much less popular as compared to standard deviation.

5.4. STANDARD DEVIATION:

Ungrouped Data

- ☆ It is defined as the square root of the mean of the squares of the deviations taken from Arithmetic mean.
- ☆ The standard deviation is denoted by 's' in case of sample and Greek letter σ(sigma) in case of population.

Let $X_1, X_2, \dots, Xn$ be n observations then standard deviation (S.D.) is given by the formula.

$$\sigma = \sqrt{\frac{\sum_{i=1}^{n}(x_i - \overline{x})^2}{n}} \; or \sqrt{\frac{\sum_{i=1}^{n} x_i^2}{n} - (\overline{x})^2} \quad \dots (1)$$

$$\overline{x} = \frac{\sum_{i=1}^{n} x_i}{n} = A.M.$$

Example:

The weights of 5 mangoes are 100, 102,118,124,126 in gms. Find the standard deviation.

Solution:

x_i	$(xi - \overline{x})^2$
100	196
102	144
118	16
124	100
126	144
Total 570	**600**

$$\overline{x} = \frac{\sum_{i=1}^{n} x_i}{n} \text{ A. M.} = 570/5 = 114$$

$$\sigma = \sqrt{\frac{\sum_{i=1}^{n}(x_i - \overline{x})^2}{n}} = \sqrt{\frac{600}{5}} = 11$$

Grouped Data

$$\sigma = \sqrt{\frac{\sum_{i=1}^{n} f_i (x_i - \overline{x})^2}{N}} or \sqrt{\frac{\sum_{i=1}^{n} f_i x_i^2}{N} - (\overline{x})^2} \quad \ldots (2)$$

$$\overline{x} = \frac{\sum_{i=1}^{n} f_i x_i}{N} = \text{A. M.}$$

Example:

The following data pertains to weight of cucumber and number of cucumbers

Weight of cucumber (x)	160	180	200	220	240
No. of cucumbers(f)	10	25	40	15	10

Solution:

	x	f	fx	$(x-\overline{x})^2$	$f\,(x-\overline{x})^2$
	160	10	1600	1444	14440
	180	25	4500	324	8100
	200	40	8000	4	160
	220	15	3300	484	7260
	240	10	2400	1764	17640
Total		**100**	**19800**		**47600**

$$\overline{x} = \frac{\sum_{i=1}^{n} f_i x_i}{N} \text{ A. M.} = 198$$

$$\sigma = \sqrt{\frac{\sum_{i=1}^{n} f_i (x_i - \overline{x})^2}{N}} = \sigma = \sqrt{\frac{47600}{100}} = 21.8174$$

Merits and Demerits of Standard Deviation

Merits

☆ It is rigidly defined and its value is always definite and based on all the observations and the actual signs of deviations are used.

- ☆ As it is based on arithmetic mean, it has all the merits of arithmetic mean.
- ☆ It is the most important and widely used measure of dispersion.
- ☆ It is possible for further algebraic treatment.
- ☆ It is less affected by the fluctuations of sampling and hence stable.
- ☆ It is the basis for measuring the coefficient of correlation and sampling.

Demerits

- ☆ It is not easy to understand and it is difficult to calculate.
- ☆ It gives more weight to extreme values because the values are squared up.
- ☆ As it is an absolute measure of variability, it cannot be used for the purpose of comparison.

5.5 VARIANCE

The square of the standard deviation is called variance and it is denoted by σ^2

Variance = $(S.D)^2$

Ungrouped Data:

$$\sigma^2 = \frac{\sum_{i=1}^{n}(x_i - \bar{x})^2}{n}$$

Grouped Data:

$$\sigma^2 = \frac{\sum_{i=1}^{n} f_i (x_i - \bar{x})^2}{N}$$

5.6 RELATIVE MEASURE OF DISPERSION

The relative measures of dispersion are used to compare the two or more data sets. Another name of Relative measure of dispersion is coefficients of dispersion.

The common relative dispersion measures involved

- ☆ Co-efficient of Range

  Co-efficient of Range $= \dfrac{X_{max} - X_{min}}{X_{max} + X_{min}}$

- ☆ Co-efficient of Quartile Deviation

  Co-efficient of Quartile Deviation = $\dfrac{Q_3 - Q_1}{Q_3 + Q_1}$

- ☆ Co-efficient of Mean Deviation

  Co-efficient of Mean Deviation = $\dfrac{Mean deviation}{Average}$

☆ Co-efficient of Variation(CV)

$$\text{Co-efficient of Variation} = \frac{\textit{Standard deviation}}{\textit{Arthmetic mean}}$$

$$\text{CV\%} = \frac{\textit{Standard deviation}}{\textit{Arthmetic mean}} \times 100$$

☆ Less CV% indicates more consistency of data

☆ More CV% indicates less consistency of data

CHAPTER-6

ELEMENTARY PROBABILITY

6.1 BASIC CONCEPTS OF PROBABILITY

Sample Space

A Set of possible outcomes of a random experiment is called a sample space. It is denoted by 'S'

Example : A set of 4 seeds are seeded in a plot, none may be fertilize, 1,2,3 or all 4 may fertilize i.e., the possible outcomes or results are {0,1,2,3,4}. The set of numbers is called a sample space.

Trial and Event

An Experiment is repeated under essentially identical (or) unique condition does not give the same results but may result in any one of the several possible outcomes. Conducting an experiment is known as a "Trial".

Every non-empty subset of a sample space of a Probability or Random experiments is known as an "Event".

Problems

- ☆ Seed fertilization – either fertilizes or does not fertilizes are events.
- ☆ In a lot of 5 seeds none may fertilize 0, 1 or 2 or 3 or 4 or all 5 may fertilize.

Exhaustive Events

The total number of possible outcomes in any trial is known as exhaustive events (or) exhaustive cases or Total number of cases.

Example:

When insecticide is applied insects may survive or die. There are two exhaustive cases namely (Survival, Death).

Favourable Events

The number of cases favourable to an event in a trial is the number of outcomes which involve the happening of the event is called as Favourable Events.

Example:

When a seed is seedling if we observe non fertilize of a seed, it is a favourable event. If we are interested in fertilize of the seed then germination is the favourable event.

Mutually Exclusive Events

Events are said to be Mutually Exclusive if the happening of any one of the event excludes or eliminates the happening of all other remaining events are called as Mutually Exclusive Events i.e., The join occurrence is not possible.

Example:

When a seed is seedling if we observe it may fertilize or may not be fertilize of a seed. Fertilize and non fertilize are Mutually Exclusive events.

Note:

- ☆ If a coin is tossed 'n' times (or) 'n' coins are tossed at a time then the sample space consists of '2^n'.
- ☆ If a dice is thrown 'n' or 'n' dice are thrown at a time then the sample space consists of '6^n'.
- ☆ If two events are mutually exclusive then the sets are disjoint i.e., $A \cap B = \emptyset$.

Some Basic Formulae:

- ☆ $A \cup A = A$; $A \cap A = A$
- ☆ $A \cup (B \cup C) = (A \cup B) \cup C$; $A \cap (B \cap C) = (A \cap B) \cap C$
- ☆ $A \cap B = B \cap A$; $A \cup B = B \cup A$
- ☆ $A \cup \emptyset = A$; $A \cap \emptyset = \emptyset$
- ☆ $A \cup A' = S$ (Sample space)
- ☆ $A \cap A' = \emptyset$
- ☆ $(A')' = A$
- ☆ $(A \cup B)' = A' \cap B'$; $(A \cap B)' = A' \cup B'$
- ☆ $A \cup (B \cap C) = (A \cup B) \cap (A \cup C)$; $A \cap (B \cup C) = (A \cap B) \cup (A \cap C)$

Mutually Independent Events

Events are said to be mutually independent, if the incidence of any one of the events does not effect the incidence of any other events.

- ☆ If A, B are two independent events then $P(A \cap B) = P(A) * P(B)$

Dependent Events

Events are said to be dependent, if the incidence of any one of the events effect the incidence of one or more events.

6.2 DEFINITION OF PROBABILITY

Mathematical (Or) Classical (Or) Apiori Probability

The word 'Probability' used commonly and several times in our daily life such as 'Probability this year we may have a good harvest, '

If an experiment is conducted and 'n 'is the total number possible cases and 'm' is the number of favourable cases of an event 'E' then the probability of an event 'E' is defined as :

$$P(E) = \frac{Number\ of\ favourable\ cases}{Total\ number\ possible\ cases} = \frac{m}{n}$$

Example : When a coin is tossed then the probability of getting a head is $P(E) = \frac{1}{2}$

6.3 AXIOMS OF PROBABILITY

The axioms of probability are

(i) Probability of positivity:

$0 \leq P(A) \leq 1$ for each event A C S

i.e., Probability always lies between 0 to 1.

(ii) Probability of certainity:

P(S) =1

i.e., Probability of a sample space is always 1.

(iii) Probability of addition:

If A and B are two mutually exclusive events then

$P(AUB) = P(A) + P(B)$ ($\because P(A \cap B) = P(\emptyset) = 0$)

Note:

☆ $P(A^1) = 1 - P(A)$, Here $A^1 = \overline{A}$

☆ $P(A^1 \cap B) = P(B) - P(A \cap B)$.

☆ P(A) is called probability of success.

☆ $P(A^1)$ is called probability of failure.

☆ P(A) = 0, the event A is called the impossible event.

6.4 PROBLEMS IN PROBABILITY

Q. What is the probability for a leap year to have 52 Mondays and 53 Sundays?

Solution:

We know that,

Leap year has 366 days ($\because \frac{366}{7} = 52$ weeks and 2 days)

The number of possibilities for 2 days are

{(Mon, Tue);(Tue, Wed);(Wed, Thur);(Thur, Fri);(Fri, Sat);(Sat, Sun);(Sun, Mon)}

Probability of getting 52 sundays in a leapyear:

Combination of (sun,Mon) &(Sat, Sun) has to be excluded because we aready have 52 sundays.

Required Probability = days without sundays/Total combination = 5/7.

Probability of getting 53 sundays in a leap year:

Required Probability = days with sundays / total combination = 2/7

$\therefore$ The Probability of getting 52 Mondays and 53 Sundays = 5/7 and 2/7

Q. Among 170 students are studying Horticulture students, 60 are studying Fruit Science, 40 are studying Vegetable Science, 70 are studying Fruit Science and Vegetable Science If a student is chosen at random, find the probability that the student (i) Studying Fruit Science or Vegetable Science (ii) Studying neither Fruit Science nor Vegetable Science.

Solution:

Given that,

Total number of students = 170.

Let 'A' be the event that, the number of students studying Fruit Science =

$\therefore n(A) = 60$

The number of students studying Vegetable Science =

$\therefore n(B) = 40$

The number of students studying both Fruit Science and Vegetable Science =

$n(A \cap B) = 70$

$\therefore$ The probability of a student studying Fruit Science = $\frac{60}{170}$

$\therefore$ The probability of a student studying Vegetable Science = $\frac{40}{170}$

$\therefore$ The probability of a studying both Fruit Science and Vegetable Science = $\frac{70}{170}$

(i) The probability of a Studying Fruit Science or Vegetable Science is

$P(AUB) = P(A) + P(B) - P(A \cap B = \frac{60}{170} + \frac{40}{170} - \frac{70}{170} = \frac{30}{170}$

$P(AUB) = \frac{3}{17}$

(ii) The probability of a Studying neither Fruit Science nor Vegetable Science is

$P(A' \cap B') = 1 - P(AUB) = 1 - \frac{3}{17} = \frac{14}{17}$

CHAPTER-7

PROBABILITY DISTRIBUTIONS

Let 'X' is a number determined by the outcome of an experiment, so 'X' is called a random variable and the set {X, P(x)} is called probability function of 'X'.

Here the table shows the survivals (S) and its probability:

Sample points	DD	SS	DS	SD
X (Number of Survivals)	0	2	1	1
P(X)	1/4	1/4	1/4	1/4

Therefore, we have

X (Number of Survivals)	0	1	2
P(X)	1/4	1/2	1/4

Discrete and Continuous Distribution Function

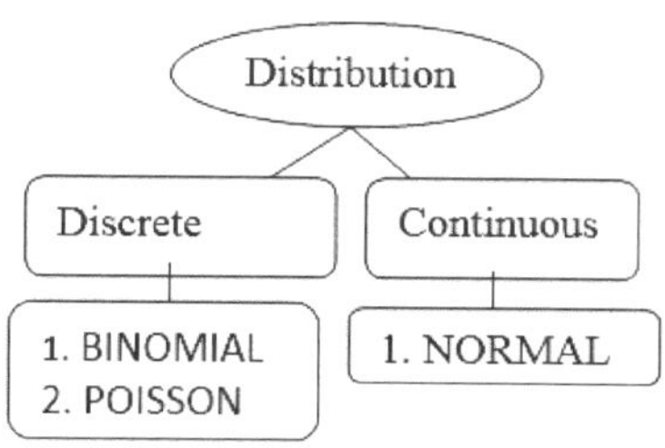

7.1 BINOMIAL DISTRIBUTION

Bernoulli Distribution

Introduced by Swiss mathematician James Bernolli (1654-1705). A random variable concentration to study on success (1's) and failures (0's) with probabilities p and q is called as bernolli variate.

Here $q = 1-p$

Example :

There are 15 seed samples, each sample consists of say 50 seeds. A sample maybe either accepted if 5 or less seeds do not sprout, otherwise rejected. There are 50 outcomes in which each sample may be either rejected or accepted. These 50 outcomes may be considered as 50 Bernolli trails.

Note:

- ☆ Probability mass function = P(X=x)
- ☆ Total probability = $\sum_{x=0}^{1} P(x) = 1 \Rightarrow p+q=1$
- ☆ $0 \le p \le 1$ and also $0 \le q \le 1$

The probability of x successes out of n trails or at least or utmost x successes out of n trails is obtained as follows.

If the probability of success by *'p'* and failure by *'q'*, then there will be ways in which x successes can occur in *'n'* trails. The probability of each combination of x successes and $(n-x)$ failures is $p^x q^{(n-x)}$, Since each trail is independent from each trail.

Then the Probability mass function of binominal distribution is given by

$$P(X=x) = \binom{n}{x} p^x q^{(n-x)};\ x = 0,1,2,\ldots n$$

7.1.1 Properties of Binomial Distribution

- ☆ n = The number trails are finite. n, p are parameters.
- ☆ Number of trails is independent.
- ☆ There are two possible outcomes i.e. success or failure
- ☆ Mean of Binomial Distribution = np and variance = npq
- ☆ If p = q = 0.5 Binominal distribution will be symmetrical. If $p \neq q$ Binominal distribution will be asymmetrical. When p<q the distribution will be positively skewed. When p > q the distribution will be negatively skewed.
- ☆ 6. When n-> ∞ and p - > 0 binominal distribution tends to be standard normal distribution.

7.1.2 Fitting of the Binomial Distribution

To estimate the parameters of Binominal distribution and test the significance of observed and expected frequencies with the help of fitting of Binominal distribution.

To calculate the expected frequency, we use

f(x) = N*P(X = x)

Here, N = Total frequency

$$P(X=x) = \binom{n}{x} p^x q^{(n-x)};\ x = 0,1,2,\ldots n$$

7.1.3 Problems

Q. A pharmaceutical lab states that a drug causes negative side effects in 6 of every 100 patients. To confirm this affirmation, another laboratory chooses 10 people at random who have consumed the drug. What is the probability of the following events?

- ☆ None of the five patients experience side effects.
- ☆ At least three experience side effects.
- ☆ What is the average number of patients that the laboratory should expect to experience side effects if they choose 100 patients at random?

Solution:

$X \sim (n, P)$

$P(X = x) = p(x) = n_{c_x}\ p^x\ q^{n-x}$ ... (1)

$X = 0,1,2\ldots n$

$q = 1-p$

Q. None of the ten patients experience side effects.

$B(100,0.06)$, $p = 0.06$, $q = 0.94$

$$P(X=x) = \binom{6}{0}(0.06)^0(0.94)^6 = 0.6898$$

2) At least three experience side effects.

$P(X \geq 3) = 1 - P(X < 3) = 1 - [p(x=0) + p(x=1) + p(x=2)]$

$$1-\left[\binom{6}{0}(0.06)^0(0.94)^6+\binom{6}{1}(0.06)^1(0.94)^5+\binom{6}{2}(0.06)^2(0.94)^4\right]$$

$$1-[0.6898+0.264+0.0422] = 0.004$$

Q. What is the average number of patients that the laboratory should expect to experience side effects if they choose 100 patients at random?

$\mu = 100*0.03 = 3$

Q) Find the binomial distribution if the mean is 15 and variance is 4

Solution:

We are given that,

Mean $= np = 15$

Variance $= npq = 3$

$$\frac{Mean}{Variance} = \frac{np}{npq} = \frac{15}{3} = 5$$

$$=> \frac{1}{q} = 5$$

=> q = 0.2

=> p = 1 – 0.2 = 0.8

=> np = 15 *0.8 = 12 = mean

=> npq = 15 *0.8 * 0.2 = 2.4 =Variance

7.2 Poisson Distribution

A discrete random 'X' is said to follow Poisson distribution it assumes only non-negative values and its probability mass function is given by

$$P(X = x) = \frac{e^{-\lambda}\lambda^{x}}{x!};\ x = 0,1,\ \ldots,;\infty;\lambda > 0, x = 0,\ 1,\ldots\infty$$

e is the Euler's number (e = 2.718)

Here λ is parameter.

7.2.1 Properties of Poisson Distribution

- ☆ Poisson distribution as a limiting case of Binominal distribution *i.e.* n - > ∞ and np =λ is finite.
- ☆ Mean = variance
- ☆ If the n - > ∞, then the Poisson distribution is approximately a normal distribution.
- ☆ In poisson distribution n - > ∞, and p - > 0
- ☆ Poisson distribution could be uni-modal or bi-modal depending upon the value of the parameter λ.
- ☆ Possion distribution is applied for rare events.

7.2.2 Fitting of the Poisson Distribution

To estimate the parameters of Poisson distribution and test the significance of observed and expected frequencies with the help of fitting of Poisson distribution.

To calculate the expected frequency, we use

f(x) = N*P(X = x)

Here, N = Total frequency

$$P(X = x) = \frac{e^{-\lambda}\lambda^{x}}{x!};\ x = 0,1,\ \ldots,\infty;\lambda > 0,$$

7.2.3 Problems

Q) i)Suppose a book of 585 pages contains 43 typographical errors. If these errors are randomly distributed throughout the book, what is the probability that 10 pages, selected at random, will be free from errors?

Solution:

Given that,

Total No. of pages = Total No. of cases = 585

Typographical errors in that pages = favorable No. of cases = 43

$$P = \frac{\text{favorable No.of cases}}{\text{Total No.of cases}} = \frac{43}{585} = 0.0735 \; ; q = 1 - p = 1 - 0.0735 = 0.9265$$

N is very large and p is very small, then it follows the Poisson distribution

n = 10, Mean = np = λ

Therefore Mean = 10*0.0735 = 0.735

The probability function of Poisson distribution is given by

$$P(x) = \frac{e^{-\lambda}\lambda^x}{x!} = \frac{e^{-0.735}0.735x}{x!}$$

Probability of free from errors = P(0) = $\frac{e^{-0.735}0.735^0}{0!} = \frac{0.4725}{1} = 0.4725$

ii) When counting red blood cells, a square grid is used, over which a drop of blood is evenly distributed. Under the microscope an average of 8 erythrocytes are observed per single square. What is the probability that exactly 5 erythrocytes are found in one square?

Solution:

Let X be a random variable follows Poisson distribution with number of erythrocytes.

Hence, Mean λ = 8 erythrocytes/single square

P(exactly 5 erythrocytes are in one square) = P(X = 5) = $\frac{e^{-\lambda}\lambda^x}{x!}$

$$= \frac{e^{-8}8^5}{5!}$$

= 0.0915

The probability that exactly 5 erythrocytes are found in one square is 0.0916. i.e there are 9.16% chances that exactly 5 erythrocytes are found in one square.

iii) Assuming one in 80 births is a case of twins, calculate the probability of 2 or more sets of twins on a day when 30 births occur.

Solution:

Let x denotes the set of twins on a day

P(twin birth) = p = 1/80 = 0.0125 and n = 30

The value of mean $\lambda = np = 30 \times 0.0125 = 0.375$

Hence, X follows poisson distribution with $p(x) = \frac{e^{-\lambda}\lambda^{x}}{x!}$

The probability is

P(2 or more) = $p(X \geq 2)$ or $1 - p(x \leq 1) = 1 - [p(x = 0) + p(x = 1)]$

$= 1 - \left[\frac{e^{-0.375}0.375^{0}}{0!} + \frac{e^{-0.375}0.375^{1}}{1!}\right] = 1 - e^{-0.375}\{1 + 0.375\}$

$= 1 - (0.6873 \text{ X } 1.375) = 0.0549$

iii) If the probability that an individual seed stays dormant is 0.001, determines the probability that out of 2,000 seeds (a) exactly 3, and (b) more than 2 individuals will stay dormant.

Solution:

Consider a 2,000 individuals seeds, n = 2000

Let X be the number of individual seed stays dormant.

Let p be the probability that an individual seed stays dormant = 0.001

and $q = 1 - p = 1 - 0.001 = 0.999$

Since n is large and p is small, Binomial Distribution approximated to poisson distribution

So, $\lambda = np = 2000 \times 0.001 = 2$

(i) Probability out of 2000, exactly 3 will suffer a bad reaction is

$$P(X = 3) = \left[\frac{e^{-2}2^{3}}{3!}\right] = 0.1804; e^{-2} = 0.1353$$

(ii) Probability out of 2000, more than 2 individuals will suffer a bad reaction

$= P(X > 2)$

$= 1 - [P(X \leq 2)]$

$= 1 - [P(x = 0) + P(x = 1) + P(x = 2)]$

$$= 1 - \left[\frac{e^{-2}2^{0}}{0!} + \frac{e^{-2}2^{1}}{1!} + \frac{e^{-2}2^{2}}{2!}\right]$$

$= 1 - e^{-2}[1 + 2 + 2]$

$= 1 - (0.1353 \text{ X } 5) = 0.3235$

7.3 Normal Distribution

Normal distribution is continuous probability distribution. The another name of the normal distribution is Laplace distribution (or) Gaussian distribution.

If 'X' be a random variable with mean μ and variance σ^2 then its probability density function is given by

$$f(x;\ \mu,\sigma^2)=\frac{1}{\sigma\sqrt{2\pi}}e^{\frac{-1}{2}(\frac{x-\mu}{\sigma})^2};-\infty<x<\infty;-\infty<\mu<\infty \text{ and } \sigma^2>0$$

7.3.1 Properties of Normal Distribution

- ☆ If $Z=\frac{x-\mu}{\sigma}$, then probability density function of normal distribution is given by

  $$f(z)=\frac{1}{\sqrt{2\pi}}e^{\frac{-1}{2}(z)^2}$$;Here μ= mean=0 and variance=σ^2=1. It called as standard normal distribution.

- ☆ It is symmetrical distribution.
- ☆ If n -> ∞ binominal and poisson distributions tends to be standard normal distribution.
- ☆ Normal curve is bell shaped curve.

7.3.2 Fitting of the Normal Distribution

To estimate the parameters of Normal distribution and test the significance of observed and expected frequencies with the help of fitting of Normal distribution.

To calculate the expected frequency, we use

E(x) = N*f(X = x)

Here, N = Total frequency

$$f(x;\ \mu,\sigma^2)=\frac{1}{\sigma\sqrt{2\pi}}e^{\frac{-1}{2}(\frac{x-\mu}{\sigma})^2};-\infty<x<\infty;-\infty<\mu<\infty \text{ and } \sigma^2>0$$

7.3.3 Problems

Q. What is the probability that a standard normal variate Z will be

- ☆ greater than 1.09
- ☆ less than -1.65

☆ lying between -1.00 and 1.96

☆ lying between 1.25 and 2.75

Solution:

(i) greater than 1.09

The total area under the curve is equal to 1 , so that the total area to the right Z = 0 is 0.5 (since the curve is symmetrical). The area between Z = 0 and 1.09 (from tables) is 0.3621

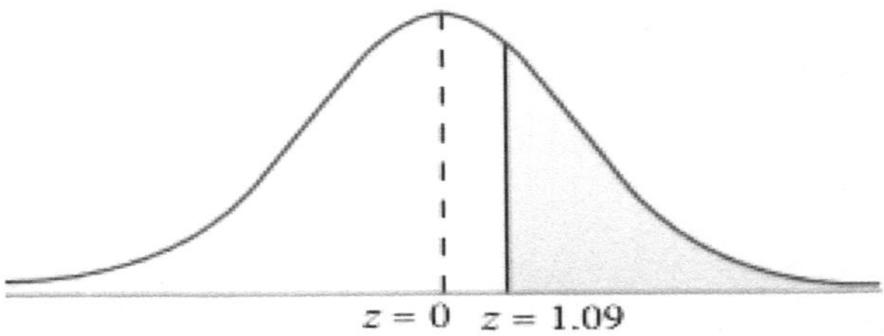

Fig 1

P(Z > 1.09) = 0.5000 - 0.3621(table value) = 0.1379

The shaded area to the right of Z = 1.09 is the probability that Z will be greater than 1.09

(ii) less than –1.65

The area between -1.65 and 0 is the same as area between 0 and 1.65. In the table the area between zero and 1.65 is 0.4505 (from the table). Since the area to the left of zero is 0.5 , P(Z< 1.65) = 0.5000 – 0.4505(table value) = 0.0495.

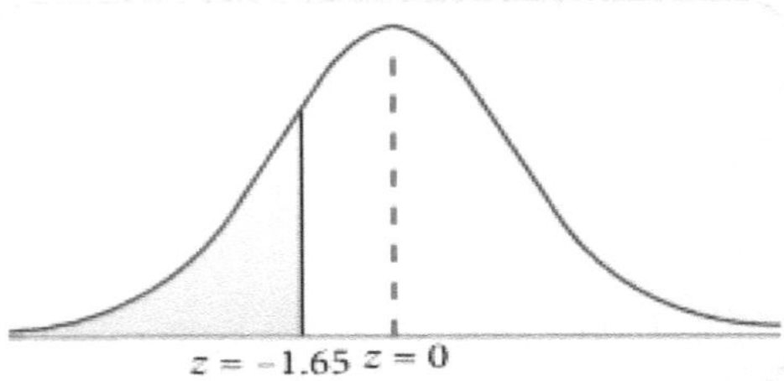

Fig 2

(iii) lying between -1.00 and 1.96

The probability that the random variable Z in between -1.00 and 1.96 is found by adding the corresponding areas :

Area between -1.00 and 1.96 = area between (-1.00 and 0) + area betwn (0 and 1.96)

P(–1.00 < Z < 1.96) = P(–1.00 < Z < 0) + P(0 < Z < 1.96) = 0.3413 + 0.4750 (by tables) = 0.8163

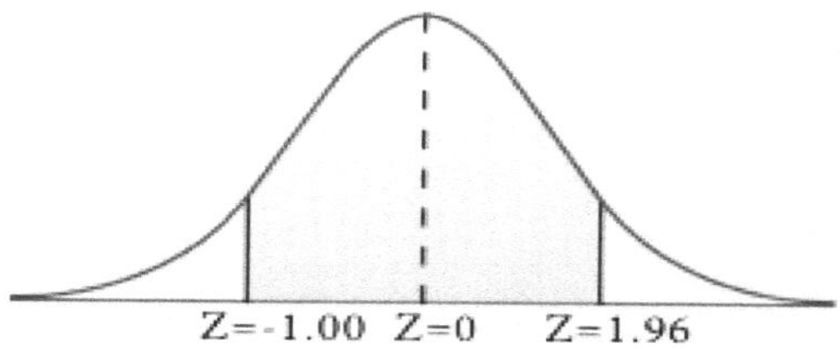

Fig 3

(iv) lying between 1.25 and 2.75

Area between Z = 1.25 and 2.75 = area between (z = 0 and z = 2.75) – area between (z=0 and z = 1.25)

P(1.25 < Z < 2.75) = P(0 < Z < 2.75) – P(0 < Z < 1.25)

= 0.4970 - 0.3944 = 0.1026

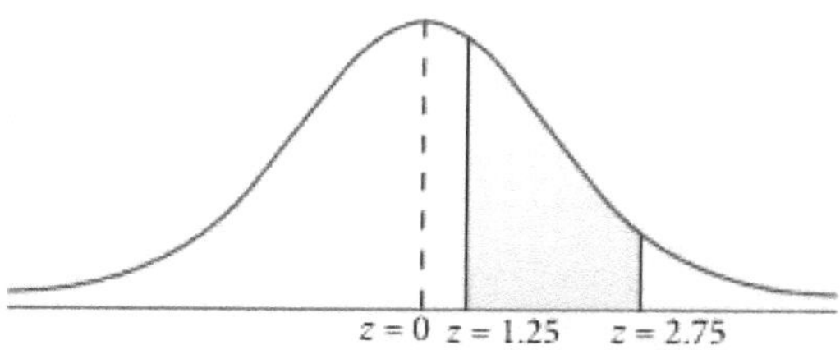

Fig 4

Q. If *X* is a normal variate with mean 30 and SD is 5. Find the probabilities that (i) 26 ≤ *X* ≤ 40 (ii) *X* > 45

Solution:

Here mean μ= 30 and standard deviation σ = 5

(i) When X = 26, Z = (X – μ) / σ = (26 – 30)/5 = –0.8

And when X = 40 , Z = [40 – 30] /5 = 2

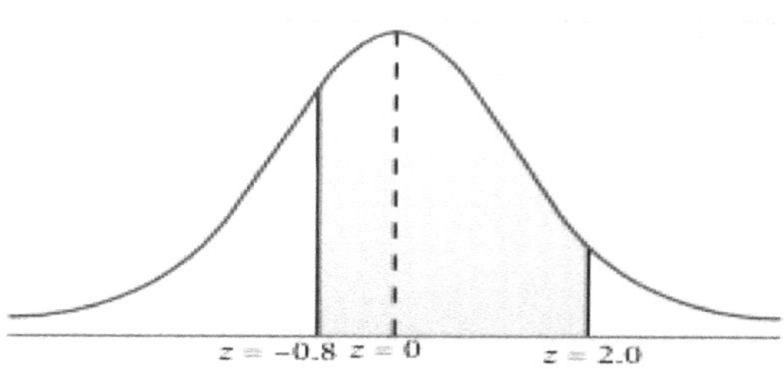

Fig1

Therefore,

$P(26<X<40)\ ) = P(-0.8 \leq Z < 2)$

$= P(-0.8 \leq Z \leq 0) + p(0 \leq Z \leq 2)$

$= P(0 \leq Z \leq 0.8) + P(0 \leq Z \leq 2)$

$= 0.2881 + 0.4772$ (By tables)

$= 0.7653$

(ii) The probability that $X \geq 45$

When $X = 45$

$Z = [X - \mu] / \sigma = [45 - 30] / 5 = 3$

$= P(X \geq 45) = P(Z \geq 3)$

$= 0.5 - 0.49865$

$= 0.00135$

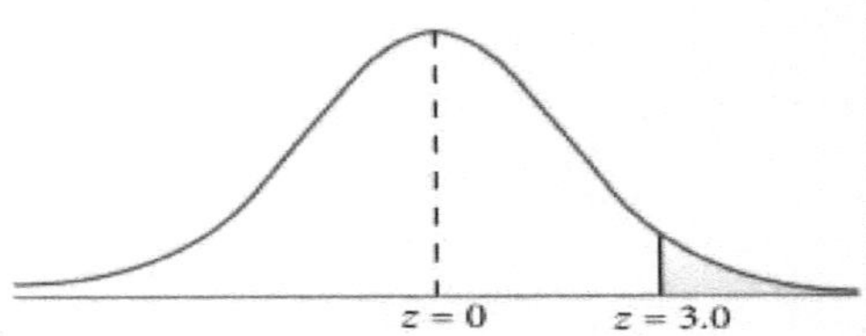

Fig 2

Q. The average daily sale of 550 branch offices was Rs.150 thousand and standard deviation is Rs. 15 thousand. Assuming the distribution to be normal, indicate how many branches have sales between

- ☆ Rs. 1,25,000 and Rs. 1, 45, 000
- ☆ Rs. 1,40,000 and Rs.1,60,000

Solution:

Given that mean μ= 150 and standard deviation $\sigma = 15$

when X = 125 thousand

$[X - \mu] / \sigma = (125\text{-}150) / 15 = -1.667$

when X = 145 thousand

$[X - \mu] / \sigma = (145\text{-}150) / 15 = -0.33$

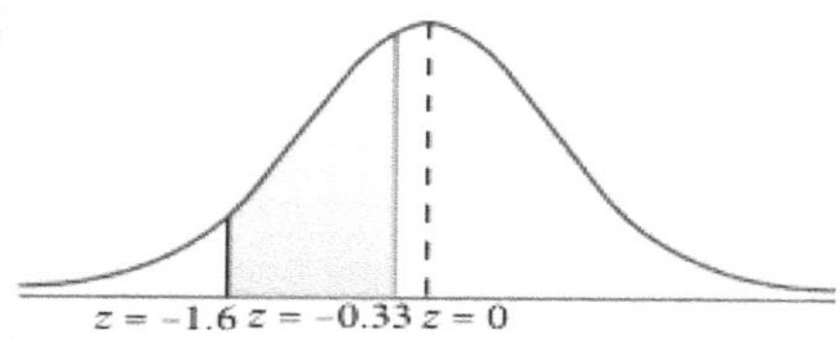

Area between Z = 0 and Z = -1.67 is 0.4525

Area between Z = 0 and Z = -0.33 is 0.1293

$P(-1.667 \leq Z \leq -0.33) = 0.4525 - 0.1293$

$= 0.3232$

Therefore the number of branches having sales between 125 thousand and 145 thousand is 550 × 0.3232 = 178

When X = 140 thousand

$Z = [X - \mu] / \sigma = [140 - 150] / 15 = -0.67$

When X = 160 thousand

$Z = [X - \mu] / \sigma = [160 - 150] / 15 = 0.67$

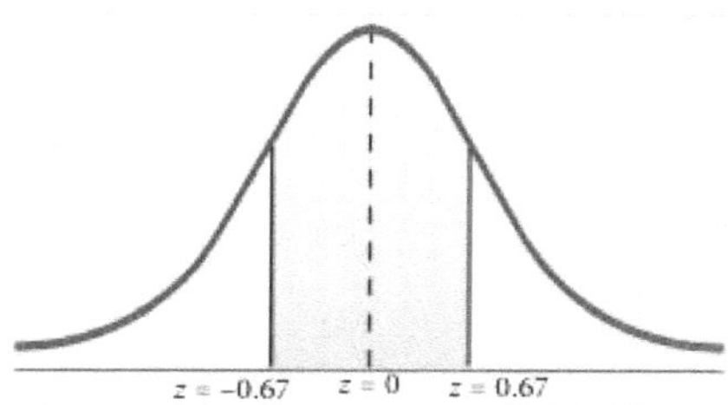

$P(-0.67 < Z < 0.67) = P(-0.67 < Z < 0) + P(0 < Z < 0.67)$

$= P(0 < Z < 0.67) + P(0 < Z < 0.67)$

$= 2\ P(0 < Z << 0.67)$

$= 2 \times 0.2486$

$= 0.4972$

Therefore, the number of branches having sales between Rs.140 thousand and Rs.160 thousand = 550 × 0.4972 = 273.

CHAPTER-8

TESTING OF HYPOTHESIS

Statistical inference is basically drawing of valid statistical conclusions about the population characteristics on the basis of a sample drawn from the population in a systematic manner. Two important problems in statistical inference are: (i) estimation and (ii) testing of hypothesis.

In many situations, one may be interested in the value of an unknown population characteristic, which is known as estimation, whereas the testing of hypothesis is one which helps to collect a representative sample and examine it to see if our hypothesis, which is a statement about population holds true. When we study the Testing of Hypothesis, first know about the following terms

8.2 CONCEPTS OF TESTING OF HYPOTHESIS

Statistical Population Or Population (Universe)

Population means aggregate of all possible units. It need not be human population. It may be population of plants, population of insects, population of fruits, etc.

Finite Population

When the number of observation can be counted and is definite, it is known as finite population

- ☆ No. of plants in a plot.
- ☆ No. of farmers in a village.
- ☆ All the fields under a specified crop.

Infinite Population

When the number of units in a population is innumerably large, that we cannot count all of them, it is known as infinite population.

☆ The plant population in a region.

☆ The population of insects in a region.

Sample

A portion or small number of unit of the total population is known as sample.

☆ All the farmers in a village(population) and a few farmers(sample)

☆ All plants in a plot is a population of plants.

☆ A small number of plants selected out of that population is a sample of plants.

Frame

A list of all units of a population is known as frame.

Parameter

A summary measure that describes any given characteristic of the population is known as parameter. Population are described in terms of certain measures like mean, standard deviation etc. These measures of the population are called parameter and are usually denoted by Greek letters. For Example , population mean is denoted by μ, standard deviation by σ and variance by σ^2 .

Statistic

A summary measure that describes the characteristic of the sample is known as statistic. Thus sample mean, sample standard deviation etc is statistic. The statistic is usually denoted by roman letter.

$\bar{x}$ - sample mean

s – Sample standard deviation

Note: The statistic is a random variable because it varies from sample to sample.

Sampling Distribution

By drawing all possible samples of same size from a population we can calculate the statistic, for Example , for all samples. Based on this we can construct a frequency distribution and the probability distribution. Such probability distribution of a statistic is known a sampling distribution of that statistic. In practice, the sampling distributions can be obtained theoretically from the properties of random samples.

Standard Error

As in the case of population distribution the characteristic of the sampling distributions are also described by some measurements like mean & standard deviation. Since a statistic is a random variable, the mean of the sampling distribution of a statistic is called the expected valued of the statistic.

The Standard deviation of the sampling distributions of the statistic is called standard error of the Statistic. The square of the standard error is known as the variance of the statistic. It may be noted that the standard deviation is for units whereas the standard error is for the statistic.

Statistical Hypothesis

☆ When the assumption or statement that occurs under certain conditions is formulated as scientific hypothesis, we can construct criteria by which a scientific hypothesis is either rejected or provisionally accepted. For this purpose, the scientific hypothesis is translated into statistical language. If the hypothesis in given in a statistical language it is called a statistical hypothesis.

For *e.g.*

The yield of a new paddy variety will be 3500 kg per hectare – scientific hypothesis. In Statistical language if may be stated as the random variable (yield of paddy) is distributed normally with mean 3500 kg/ha.

☆ A hypothesis is a statement about the population developed for the purpose of testing is called Statistical hypothesis. Whereas the form of the probability distribution is known then any statement which refers to the values of the unknown parameters of that distribution is called a parametric hypothesis.

Further, a hypothesis is defined in two ways:

(i) Simple hypothesis, and

(ii) Composite hypothesis.

For Example , $x_{1,}$ $x_{2\,...}$ x_n are random samples from a normal distribution with mean μ and variance σ^2 which is known.

Simple Hypothesis

A parametric hypothesis which specifies completely all the parameters in a probability distribution is called a simple hypothesis.

$H_0 : \mu = \mu_0$ is a simple hypothesis

Composite Hypothesis

Which fails to completely specify the probability distribution is called a composite hypothesis.

Whereas the hypothesis $\mu > \mu_0$ or $\mu < \mu_0$ are the composite hypotheses.

The Hypothesis testing is a stratagem which evolves on sample evidence and probability theory, used to determine whether the hypothesis is a rational statement and should not be rejected, or is irrational and should be rejected.

Null Hypothesis - H_o

- ☆ Consider for Example, the hypothesis may be put in a form 'coconut variety A will give the same yield per hectare as that of variety B' or there is no difference between the average yields of coconut varieties A and B. These hypotheses are in definite terms. Thus these hypothesis form a basis to work with. Such a working hypothesis in known as null hypothesis. It is called null hypothesis because if nullities the original hypothesis, that variety A will give more yield than variety B.
- ☆ The null hypothesis is stated as 'there is no difference between the effect of two treatments or there is no association between two attributes (ie) the two attributes are independent. Null hypothesis is denoted by Ho.

eg:-

- ☆ There is no significant difference between the yields of two coconut varieties (or) they give same yield per unit area. Symbolically, Ho: $\mu_1 = \mu_2$.

Alternative Hypothesis (H_1)

When the original hypothesis is $\mu_1 > \mu_2$ stated as an alternative to the null hypothesis is known as alternative hypothesis. Any hypothesis which is complementary to null hypothesis is called alternative hypothesis, usually denoted by H_1.

eg:-

There is a significance difference between the yields of two coconut varieties.

☆ Symbolically,

H_1: $\mu_1 \neq \mu_2$ (two sided or directionless alternative)

If the statement is that A gives significantly less yield than B (or) A gives significantly more

yield than B. Symbolically,

$H_{1:}$ $\mu_1 < \mu_2$ (one sided alternative-left tailed)

H_1: $\mu_1 > \mu_2$ (one sided alternative-right tailed)

Sampling Error

From sample data, the statistic is computed and the parameter is estimated through the statistic. The difference between the parameter and the statistic is known as the sampling error.

Decision Errors

By performing a test we make a decision on the hypothesis by accepting or rejecting the null hypothesis Ho. In the process we may make a correct decision on Ho or commit one of two kinds of error.

- ☆ We may reject Ho based on sample data when in fact it is true. This error in decisions is known as Type I error.
- ☆ We may accept Ho based on sample data when in fact it is not true. It is known as Type II error.

	Accept Ho	Reject Ho
Ho is true	Correct Decision	Type I error
Ho is false	Type II error	Correct Decision

The relationship between type I & type II errors is that if one increases the other will decrease.

- ☆ The probability of type I error is denoted by α. The probability of type II error is denoted by β.
- ☆ The correct decision of rejecting the null hypothesis when it is false is known as the power of the test. The probability of the power is given by $1-\beta$.

Critical Region

The testing of statistical hypothesis involves the choice of a region on the sampling distribution of statistic. If the statistic falls within this region, the null hypothesis is rejected: otherwise it is accepted. This region is called critical region.

Let the null hypothesis be Ho: $\mu_1 = \mu_2$ and its alternative be H_1: $\mu_1 \neq \mu_2$. Suppose Ho is true. Based on sample data it may be observed that statistic $(\overline{X}_1 - \overline{X}_2)$ follows a normal distribution given by

$$Z = \frac{(\overline{X}_1 - \overline{X}_2) - (\mu_1 - \mu_2)}{SE(\overline{X}_1 - \overline{X}_2)}$$

We know that 95% values of the statistic from repeated samples will fall in the range $(\overline{X}_1 - \overline{X}_2)$ ±1.96 times SE $(\overline{X}_1 - \overline{X}_2)$.

This is represented by a diagram.

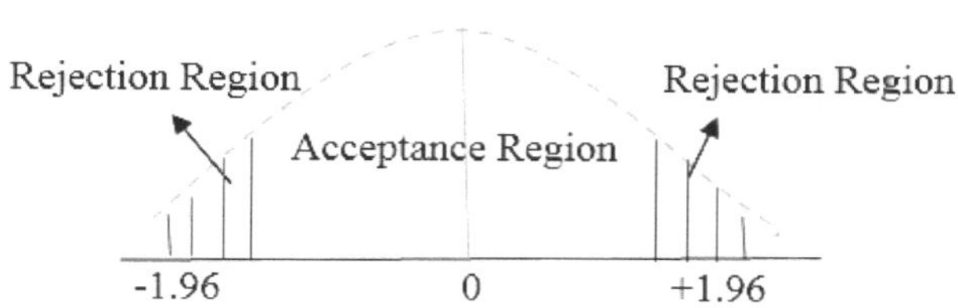

The border line value ±1.96 is the critical value or tabular value of Z.

- ☆ The area beyond the critical values (shaded area) is known as critical region or region of rejection. The remaining area is known as region of acceptance.
- ☆ If the statistic falls in the critical region we reject the null hypothesis and, if it falls in the region of acceptance we accept the null hypothesis.
- ☆ In other words if the calculated value of a test statistic (Z, t, χ^2 etc) is more than the critical value in magnitude it is said to be significant and we reject Ho and otherwise we accept Ho. The critical values for the t and χ^2 are given in the form of ready made tables. Since the critical values are given in the form of table it is commonly referred as table value. The table value depends on the level of significance and degrees of freedom.
- ☆ Z cal < Z tab If We accept the Ho and conclude that there is no significant difference between the means.

Test Statistic

The sampling distribution of a statistic like Z, t, & χ^2 are known as test statistic. Generally, in case of quantitative data

$$\textit{Test statistic} = \frac{\textit{Statistic} - \textit{Parameter}}{\textit{Standard Error}\left(\textit{Statistic}\right)}$$

Note

The choice of the test statistic depends on the nature of the variable (i.e) qualitative or quantitative, the statistic involved (i.e) mean or variance and the sample size, (i.e) large or small.

Level of Significance

The probability that the statistic will fall in the critical region is α. This α is nothing but the probability of committing type I error. Technically the probability of committing type I error is known as level of Significance.

Degrees of Freedom

- ☆ The number of degrees of freedom is the number of observations that are free to vary after certain restriction have been placed on the data. If there are n observations in the sample, for each restriction imposed upon the original observation the number of degrees of freedom is reduced by one.
- ☆ The number of independent variables which make up the statistic is known as the degrees of freedom.

One and Two Tailed Test

The nature of the alternative hypothesis determines the position of the critical region. For Example , if H_1 is $\mu_1 \neq \mu_2$ it does not show the direction and hence the critical region falls on either end of the sampling distribution. If H_1 is $\mu_1 < \mu_2$ or $\mu 1 > \mu_2$ the direction is known. In the first case the critical region falls on the left of the distribution whereas in the second case it falls on the right side.

One tailed test – When the critical region falls on one end of the sampling distribution, it is called one tailed test.

Two tailed test – When the critical region falls on either end of the sampling distribution, it is called two tailed test.

For Example, consider the mean yield of new coconut variety(μ_1)is compared with that of a ruling variety (μ_2). Unless the new variety is more promising that the ruling variety in terms of yield we are not going to accept the new variety.

- ☆ In this case $H_1: \mu_1 > \mu_2$ for which one tailed test is used. If both the varieties are new our interest will be to choose the best of the two.
- ☆ In this case $H_1: \mu_1 \neq \mu_2$ for which we use two tailed test.

Steps In Testing Of Hypothesis

The process of testing a hypothesis involves following steps.

Step1 : Formulation of null & alternative hypothesis.

Step 2 : Specification of level of significance.

Step 3 : Selection of test statistic and its computation.

Step 4 : Finding out the critical value from tables using the level of significance, sampling distribution and its degrees of freedom.

Step 5 : Determination of the significance of the test statistic.

Step 6 : Decision about the null hypothesis based on the significance of the test statistic.

Step 7 : Writing the conclusion in such a way that it answers the question on hand.

Sample Theory

There are two Sample theories to study the samples drawn from population.

1. Large Sample Theory.
2. Small Sample Theory.

Large Sample Theory

The sample size n is greater than 30 ($n \geq 30$) it is known as large sample. For large samples the sampling distributions of statistic are normal(Z test). A study of sampling distribution of statistic for large sample is known as large sample theory.

Large Sample Test

Large sample test are

- ☆ Sampling from attributes
- ☆ Sampling from variables

Sampling from Attributes

There are two types of test for attributes

- ☆ Test for single proportion
- ☆ Test for equality of two proportions

Small Sample Theory

If the sample size n is less than 30 (n<30), it is known as small sample. For small samples the sampling distributions are t, F and χ^2 distribution. A study of sampling distributions for small samples is known as small sample theory.

Small Sample Test

Small sample test are

- ☆ t- test for Single mean and two means
- ☆ F- test for variances
- ☆ Chi-Square test

Test of Significance Small Tests

Based on the sampling error the sampling distributions are derived. The observed results are then compared with the expected results on the basis of sampling distribution. If the difference between the observed and expected results is more than specified quantity of the standard error of the statistic, it is said to be significant at a specified probability level. The process up to this stage is known as test of significance. The theory of test of significance consists of various test statistic. The theory had been developed under two broad heading :

- ☆ Test of significance for large sample

  Large sample test or Asymptotic test or Z test (n≥30)
- ☆ Test of significance for small samples(n<30)

  Small sample test or Exact test-t, F and χ^2.

It may be noted that small sample tests can be used in case of large samples also.

8.3 LARGE SAMPLE TESTS

Z Test for Single and Two Means

Introduction:

- ☆ In this section we describe and demonstrate the procedure for conducting a test of hypotheses about the mean of a population in the case that the sample size n is at least 30. The Central Limit Theorem states that $\overline{X}$ is approximately normally distributed, and has mean μ and standard deviation, $\frac{\sigma}{\sqrt{n}}$' where μ and σ are the mean and the standard deviation of the population. Test the significance of Large sample it having following test procedure
  - ☆ Testing of Hypothesis
  - ☆ Test statistic
  - ☆ Inference
- ☆ In case of test statistic Z, the expected value is

For Two Tailed Test

$Z\alpha = 1.96$ at 5% level

$= 2.58$ at 1% level

For one tailed test

$Z\alpha = 1.65$ at 5% level

$= 2.33$ at 1% level

- ☆ The level of significance may be fixed at either 5% or 1%

Sampling from Variable

In sampling for variables, the test are as follows

- ☆ Test for single Mean
- ☆ Test for single Standard Deviation
- ☆ Test for equality of two Means
- ☆ Test for equality of two Standard Deviation

Test Of Significance for Single Mean

In a sample of large size n, we examine whether the sample would have come from a population having a specified mean

☆ **Null Hypothesis (Ho)**

Ho: There is no significance difference between the population mean and specified value ie., $\mu=\mu o$

☆ Alternative Hypothesis (H_1)

H1 : There is significance difference between the population mean and specified value

ie., $\mu \neq \mu o$ (two-tailed)

or $\mu > \mu o$ (Right tailed)

or $\mu < \mu o$ (Left tailed)

☆ Test Statistic

When population variance is not known, it may be replaced by its estimate

$$Z = \frac{|\bar{x} - \mu_0|}{\frac{\sigma}{\sqrt{n}}}$$

S.E $\overline{X}$ = Standard error of = $\sigma/\sqrt{n}$

$\mu_0 \rightarrow$ Specified value

$\sigma \rightarrow$ Population standard deviation

$\overline{X} \rightarrow$ Sample mean

$\frac{|\bar{x} - \mu_0|}{\frac{\sigma}{\sqrt{n}}}$ Calculated Value

Z(α%) Table Value at α% Level of Significance(LoS).

☆ Level of Significance

The level may be fixed at either 10% or 5% or 1%

☆ Expected Value

The expected value is given by

Two tailed test

Z α = 1.96 at 5% level

2.58 at 1% level

One tailed test

Z α = 1.65 at 5% level

2.33 at 1% level

6. Inference

If the calculated value of the test statistic Z exceeds the table value Z we may reject the Null Hypothesis Ho otherwise accept it.

Test of Significance For Two Means

Given two sets of sample data of large size n_1 and n_2 from variables. We may examine whether the two samples come from the populations having the same mean. We may proceed as follows:

☆ Setting of Hypothesis

Null Hypothesis (Ho)

Ho: There is no significance difference between the two population means ie., $\mu_1 = \mu_2$

Alternative Hypothesis (H_1)

H_1:There is significance difference between the two population means $\mu_1 \neq \mu_2$

☆ Test Statistic

Case i: When the population variances are known and unequal (i.e) $\sigma_1^2 \neq \sigma_2^2$

$$Z = \frac{|\bar{x}_1 - \bar{x}_2|}{\sqrt{\frac{\sigma_1^2}{n_1} + \frac{\sigma_2^2}{n_2}}} \sim Z(\alpha\%)$$

$$\bar{x}_1 = \frac{\sum_{i=1}^{n} x_1}{n_1}, \bar{x}_2 = \frac{\sum_{i=1}^{n} x_2}{n_2} \text{ and}$$

$$\sigma_1^2 = \frac{\sum_{i=1}^{n}(x_1 - \bar{x}_1)^2}{n_1}; \sigma_2^2 = \frac{\sum_{i=1}^{n}(x_2 - \bar{x}_2)^2}{n_2}$$

$$\frac{|\bar{x}_1 - \bar{x}_2|}{\sqrt{\frac{\sigma_1^2}{n_1} + \frac{\sigma_2^2}{n_2}}} \rightarrow \text{Calculated Value}$$

Z (α%) →Table Value with at α% Level of Significance(LoS).

The level may be fixed at either 5% or 1%.

Case ii : When the population variances are known and equal (i.e) $\sigma_1^2 = \sigma_2^2$

$$Z = \frac{|\overline{x_1} - \overline{x_2}|}{\sqrt{\sigma^2\left(\frac{1}{n_1} + \frac{1}{n_2}\right)}}$$

$$\bar{x}_1 = \frac{\Sigma x_{1i}}{n_1} \qquad \bar{x}_2 = \frac{\Sigma x_{2i}}{n_2}$$

$$\frac{|\overline{x_1} - \overline{x_2}|}{\sqrt{\sigma^2\left(\frac{1}{n_1} + \frac{1}{n_2}\right)}} \rightarrow \textit{Calculated value}$$

$Z(\alpha\%)$ Table Value at $\alpha\%$ Level of Significance(LoS).

The level may be fixed at either 5% or 1%.

3. Inference:

Cal Z > Table value of Z, Reject H_0 otherwise Accept H_0.

Note: Test the equality of variances using F-test

8.3 SMALL SAMPLE TEST

Student's t-Distribution

- ☆ This fact was brought out by Sir William Gossest and Prof. R.A Fisher. Sir William Gossest published his discovery in 1905 under the pen name Student and later on developed and extended by Prof. R.A Fisher. He gave a test known as t-test.
- ☆ In small samples drawn from a normal population, the ratio of the difference between the sample and population means to its estimated standard error follows a distribution known as t-distribution.
- ☆ We have,
  - ☆ Single mean for t-test
  - ☆ Two independent means for t-test.
  - ☆ Paired t-test

Applications (Or) Uses

- ☆ To test the single mean in single sample case.
- ☆ To test the equality of two means in double sample case.
  - ☆ Independent samples (Independent t test)
  - ☆ Dependent samples (Paired t test)
- ☆ To test the significance of observed correlation coefficient.
- ☆ To test the significance of observed partial correlation coefficient.
- ☆ To test the significance of observed regression coefficient.

Single Mean for t-Test

Let X be random variable of size 'n' drawn from population of size N with Population mean μ and variance σ^2, When σ is unknown. Then the following steps are used to construct the Single mean for t-test.

Step 1: Setting of Hypothesis

H_0: $\mu = \mu o$ (i.e) There is no significance difference between the population mean and the specified value

H_1: $\mu \neq \mu o$ (or $\mu > \mu o$ or $\mu < \mu o$)

e.g., There is significance difference between the population mean and the specified value.

Step 2: Test statistic

$$t = \frac{|\bar{x} - \mu o|}{S/\sqrt{n}} \sim t_{n-1}(\alpha\%)$$

$$\frac{|\bar{x} - \mu o|}{S/\sqrt{n}} \rightarrow \text{Calculated Value}\ ; S^2 = \frac{\sum_{i=1}^{n} (x-\bar{x})^2}{n-1}$$

t_{n-1} ($\alpha\%$) $\rightarrow$ Table Value with (n-1) Degrees of freedom at $\alpha\%$ Level of Significance(LoS). The level may be fixed at either 5% or 1%.

Step 3: Statistical Inference

If t_{Cal} < ttab we accept the null hypothesis H_0.

We conclude that there is no significant difference population mean and specified value (or) if t_{Cal} >= t-tab we reject the null hypothesis H_0. (ie) we accept the alternative hypothesis and conclude that there is significant difference between the population mean and the specified value.

Example :

Based on field experiments, a new variety of green gram is expected to given a yield of 12.0 quintals per hectare. The variety was tested on 10 randomly selected farmer's fields. The yield (quintals/ hectare) were recorded as 14.3,12.6,13.7,10.9,13 .7,12.0,11.4,12.0,12.6,13.1. Do the results conform to the expectation?

Solution:

Step1: Setting of Hypothesis

Null hypothesis H0: μ=12.0

(i.e) the average yield of the new variety of green gram is 12.0 quintals/ hectare.

Alternative Hypothesis: H1:$\mu \neq 12.0$

(i.e) the average yield is not 12.0 quintals/hectare, it may be less or more than 12 quintals/ hectare

Level of significance: 5 % or 1%

Step 2: Test statistic

$$t = \frac{|\bar{x} - \mu o|}{S/\sqrt{n}} \sim t_{n-1}^{\alpha\%}$$

$$\bar{x} = \frac{\sum_{i=1}^{n} x}{n} = 12.63; S^2 = \frac{\sum_{i=1}^{n} (x - \bar{x})^2}{n-1} = 1.1778;\ n = 10$$

Calculated value t = 1.836

Table value for t corresponding to 5% level of significance and 9 d.f. is 2.262 (two tailed test)

Step 3: Statistical Inference

Here, tcal < ttab

We accept the null hypothesis H0

We conclude that the new variety of green gram will give an average yield of 12 quintals/hectare.

Test for Equality of Two Means (Independent Samples)

Given two sets of sample observation $x_{11}, x_{12}, x_{13} \ldots x_{1n1}$, and $x_{21}, x_{22}, x_{23} \ldots x_{2n2}$ of sizes n_1 and n_2 respectively from the normal population.

Case (i) Variances are equal

Step 1: Setting of Hypothesis

H_0: $\mu_1 = \mu_2$ (i.e) There is no significance difference between the two population means

H_1: $\mu_1 \neq \mu_2$ (or $\mu_1 > \mu 2$ or $\mu_1 < \mu 2$)

ie., There is significance difference between the two population means

Step 2: Test Statistic

$$t = \frac{|\bar{x}_1 - \bar{x}_2|}{S\sqrt{\frac{1}{n_1} + \frac{1}{n_2}}} \sim t_{n_2+n_2-2}(\alpha\%)$$

$$\bar{x}_1 = \frac{\sum_{i=1}^{n} x_1}{n_1}, \ \bar{x}_2 = \frac{\sum_{i=1}^{n} x_2}{n_2} \ and \ S^2 = \frac{(n_1 - 1)S_1^2 + (n_2 - 1)S_2^2}{n_1 + n_2 - 2}$$

$$S_1^2 = \frac{\sum_{i=1}^{n} (x_1 - \bar{x}_1)^2}{n_1 - 1}; \ S_2^2 = \frac{\sum_{i=1}^{n} (x_2 - \bar{x}_2)^2}{n_2 - 1}$$

$$\frac{|\bar{x}_1 - \bar{x}_2|}{S\sqrt{\frac{1}{n_1} + \frac{1}{n_2}}} \rightarrow Calculated\ Value$$

$t_{n_1+n_2-2}$ $(\alpha\%)$ is table value of t with ($n_1 + n_2$-2) degrees of freedom at $\alpha\%$ Level of Significance(LoS).

The level may be fixed at either 5% or 1%.

Step 3: Statistical Inference

If t_{Cal} < ttab we accept the null hypothesis H_0.

We conclude that there is no significant difference between the two population means. (or) if tCal > ttab we reject the null hypothesis H_0, (ie) we accept the alternative hypothesis and conclude that there is significant difference between the two population means.

Case (ii) **Variances are unequal and $n_1 = n_2$**

Step1: Setting of Hypothesis

H_0: $\mu_1 = \mu_2$ (i.e) There is no significance difference between the two population means

H_1: $\mu_1 \neq \mu_2$ (or $\mu_1 > \mu_2$ or $\mu_1 < \mu_2$)

ie., There is significance difference between the two population means

Step 2: Test Statistic

$$t = \frac{|\overline{x_1} - \overline{x_2}|}{\sqrt{\frac{s_1^2}{n_1} + \frac{s_2^2}{n_2}}}$$

$$\bar{x}_1 = \frac{\Sigma x_{1i}}{n_1} \qquad \bar{x}_2 = \frac{\Sigma x_{2i}}{n_2}$$

$$s_1^2 = \frac{\Sigma(x_{1i}-\overline{x_1})^2}{n_1-1} \quad s_2^2 = \frac{\Sigma(x_{2i}-\overline{x_2})^2}{n_2-1}$$

$$\frac{|\overline{x_1} - \overline{x_2}|}{\sqrt{\frac{s_1^2}{n_1} + \frac{s_2^2}{n_2}}} \dashrightarrow \textit{Calculated value}$$

$t_{n_1+n_2-2}$

Degrees of freedom at α% Level of Significance (LoS).

The level may be fixed at either 5% or 1%.

Step 3: Statistical Inference

If t_{Cal} < ttab we accept the null hypothesis H_0.

We conclude that there is no significant difference between two population means (or)

if t_{Cal} > ttab we reject the null hypothesis H_0. (ie) we accept the alternative hypothesis and conclude that there is significant difference between two population means.

Paired t- Test (or) Dependant Samples

- ☆ In the t-test for difference between two means, the two samples were independent of each other. Let us now take particular situations where the samples are not independent.
- ☆ In agricultural experiments it may not be possible to get required number of homogeneous experimental units.
- ☆ In some other situations two observations may be taken on the same experimental unit.

- ☆ For Example , the soil properties before and after the application of industrial effluents may be observed on number of plots. This will result in paired observation. In such situations we apply paired t test. Suppose the observation before treatment is denoted by x and the observation after treatment is denoted by y. for each experimental unit we get a pair of observation(x,y).
- ☆ In case of n experimental units we get n pairs of observations : (x_1, y_1), $(x_2,y_2)\ldots(x_n y_n)$. In order to apply the paired t test we find out the differences $(x_1 - y_1)$, $(x_2- y2)$,..,$(x_1 - y_1)$ and denote them as $d_1, d_2,\ldots,d_n$. Now $d_1, d_2\ldots$ form a sample. The following steps are used to construct the Paired t-test

Step 1 : Setting of Hypothesis

Ho: There is no significant difference between two pair of population means

H1: there is significant difference between two pair of population means

Step 2 : Test statistic

We apply the t test procedure for one sample (i.e),

$$t = \frac{|\bar{d}|}{S/\sqrt{n}} \sim t_{n-1}(\alpha\%)$$

$$\frac{|\bar{d}|}{S/\sqrt{n}} \rightarrow \text{Calculated Value } \bar{d} = \frac{\sum_{i=1}^{n} d_i}{n};\ S^2 = \frac{\sum_{i=1}^{n}(d_i - \bar{d})^2}{n-1}$$

t_{n-1} $(\alpha\%)$ Table Value with (n-1) Degrees of freedom at $\alpha\%$ Level of Significance(LoS). The level may be fixed at either 5% or 1%.

Step 3: Statistical Inference

If t_{Cal} < ttab we accept the null hypothesis H_0.

We conclude that there is no significant difference between two pair of population mean (or)

if t_{Cal} > ttab we reject the null hypothesis i.e. we accept the alternative hypothesis and conclude that there is significant difference between two pair of population mean.

F-Test for Variances

In case of two samples the equality of population variances has to be tested by using F-test. The following steps are used to construct F-test.

Step1 : Setting of Hypothesis

H0: $\sigma^2_1 = \sigma^2_2$ i.e., There is no significance difference between two population variances

H1: $\sigma^2_1 \neq \sigma^2_2$ i.e., there is significance difference between two population variances

Step 2 : Test Statistic

$$F= \frac{S_1^2}{S_2^2} \sim F_{(n_1-1,n_2-1)}(\alpha\%)(if\ S_1^2 > S_2^2)$$

Where S_1^2 is the variance of the first sample whose size is n_1.

S_1^2 is the variance of the second sample whose size is n_2.

$\frac{S_1^2}{S_2^2}$ Calculated Value

Table value of 'F' with at Level of Significance

It may be noted that the numerator is always the greater variance

Step3 : Statistical Inference

$F_{Cal} < Ftab$

We accept the null hypothesis H_0. *i.e.* the variances are equal otherwise the variances are unequal.

CHI SQUARE (χ^2)

χ^2- test

In case of attributes we cannot employ the parametric tests such as F and t. Instead we have to apply χ^2test. When we want to test whether a set of observed values are in agreement with those expected on the basis of some theories or hypothesis. The χ^2 statistic provides a measure of agreement between such observed and expected frequencies. The χ^2 test has a number of applications. It is used to

- ☆ Test the independence of attributes
- ☆ Test the goodness of fit
- ☆ Test the homogeneity of variances
- ☆ Test the homogeneity of correlation coefficients
- ☆ Test the equality of several proportions.

A very powerful test for testing the significance of the discrepancy between theory and experiment was given by Prof. Karl Pearson in 1900 and is known as "chi-square test of goodness of fit".

Conditions for the Validity of χ^2-Test

χ^2-test is an approximate test for large values of 'n' for the validity of -test of goodness of fit between theory and experiment, the following conditions must be satisfied.

- ☆ The sample observations should be independent.
- ☆ Constraints on the cell frequency, if any, should be linear. Example : =

$$\sum_{i=1}^{n} O_i = \sum_{i=1}^{n} E_i$$

- ☆ N, the total frequency should be reasonably large, say greater then (>) 50.
- ☆ No theoretical cell frequency should be less than (<)5. If any theoretical cell frequency is <5, then for the application of - test, it is pooled with the preceding or succeeding frequency so that the pooled frequency is more than 5 and finally adjust for degree's of freedom lost in pooling.

If 0i, (i=1, 2,…..,n) is a set of observed (experimental frequencies) and Ei (i=1,2,…..,n) is the corresponding set of expected (theoretical or hypothetical) frequencies. To construct the Chi-Square the by using the following steps are used.

Step1: Setting of Hypothesis

H_0: $O_i = E_i$ *i.e.*, there is no significance difference between observed and expected frequencies

H_1: $O_i \neq E_i$ *i.e.*, there is significance difference between observed and expected frequencies

Step 2: Test Statistic

$$\chi^2 = \sum_{i=1}^{n} \frac{(O_i - E_i)^2}{E_i} \sim \chi^2_{(n-1)}(\alpha\%)$$

$$\sum_{i=1}^{n} \frac{(O_i - E_i)^2}{E_i} \rightarrow \text{Calculated Value}$$

$\chi^2_{n-1}(\alpha\%) \rightarrow$ Table value of χ^2 with (n−1) d.f. at α% Level of Significance

Step 3: Statistical Inference

$\chi^2 Cal < \chi^2_{(n-1)}(\alpha\%)$ We Accept H_0. Otherwise Reject H_0.

Example 1

The number of yeast cells counted in a haemocytometer is compared to the theoretical value is given below. Does the experimental result support the theory?

No. of Yeast cells in the square	Observed Frequency	Expected Frequency
0	103	106
1	143	140
2	98	93
3	42	41
4	8	14
5	6	5

H_0: The experimental results support the theory

H_1: The experimental results do not support the theory.

Level of significance = 5%

Test Statistic: $\chi^2 = \sum_{i=1}^{n} \frac{(O_i - E_i)^2}{E_i} \sim \chi^2_{(n-1)} (\alpha\%)$

O_i	E_i	$O_i - E_i$	$(O_i - E_i)^2$	$(O_i - E_i)^2 / E_i$
103	106	-3	9	0.0849
143	140	2	4	0.0284
98	93	5	25	0.2688
42	41	1	1	0.0244
8	14	-6	36	2.5714
6	5	1	1	0.2
Total= 400	**Total= 400**			**Total= 3.1779**

$$\chi^2 = \sum_{i=1}^{n} \frac{(O_i - E_i)^2}{E_i} = 3.1779$$

Table value $\chi^2 = 11.070$ (at 5% LoS)

Inference

$\chi^2 cal < \chi^2 tab$

We accept the null hypothesis. (i.e) there is a good correspondence between theory and experiment.

χ^2 Test for Independence Of Attributes

At times we may consider two characteristics on attributes simultaneously. Our interest will be to test the association between these two attributes.

Example:

An entomologist may be interested to know the effectiveness of different concentrations of the chemical in killing the insects. The concentrations of chemical form one attribute. The state of insects 'killed & not killed' forms another attribute. The result of this experiment can be arranged in the form of a contingency table. In general one attribute may be divided into m classes as $A_1, A_2, \ldots A_m$ and the other attribute may be divided into n classes as $B_1, B_2, \ldots\ldots B_n$. Then the contingency table will have m x n cells. It is termed as m x n contingency table.

B \ A	A_1	A_2	...	A_m	Row total
B_1	O_{11}	O_{12}	...	$O1_m$	r_1
B_2	O_{21}	O_{22}	...	$O2_m$	r_2
. . .	. . .	. . .	...	. . .	. . .
B_n	On_1	On_2	...	On_m	r_n
Col. Total	C_1	C_2	...	C_m	Grand Total

O_{ij}s are observed frequencies.

E_{ij} The expected frequencies corresponding to O_{ij} is calculated as

(Row total * Col total)/ Grand Total

$$\chi^2 = \sum_{i=1}^{n} \frac{(O_{ij} - E_{ij})^2}{E_{ij}} \sim \chi^2_{(n-1)(m-1)}(\alpha\%)$$

where

Oij – observed frequencies

Eij – Expected frequencies

n= number of rows

m= number of columns

It can be verified that = $\sum_{i=1}^{n} O_{ij} = \sum_{i=1}^{n} E_{ij}$

This X^2 is distributed as X^2 with (n-1) (m-1) d.f.

2 X 2 χ^2 – Contingency Table

When the number of rows and number of columns are equal to 2 it is termed as 2 x 2 contingency table. It will be in the following form

	Attribute B			
Attribute A		B_1	B_2	Row Total
	A_1	a	b	$R_1 = a + b$
	A_2	c	d	$R_2 = c + d$
	Col total	$C_1 = a + c$	$C_2 = b + d$	N = a +b + c + d

Where a, b, c and d are cell frequencies C_1 and C_2 are column totals, R_1 and R_2 are row totals and N is the total number of observations.

In case of 2 x 2 contingency table can be directly found using the short cut formula, 's are observed frequencies. i= 1 to 2 and j =1 to 2, n=2 and m=2

Eij The expected frequencies corresponding to *Oij* is calculated as

(Row total * Col total)/ Grand Total

$$\chi^2 = \sum_{i=1}^{2} \frac{(O_{ij} - E_{ij})^2}{E_{ij}} \sim \chi^2_{(n-1)(m-1)}(\alpha\%)$$

Where, *Oij* – observed frequencies and *Eij* – Expected frequencies

CHAPTER-9

CORRELATION AND REGRESSION

9.1 KARL PEARSON CORRELATION

- ☆ The term correlation refers to the degree of relationship between two or more variables.
- ☆ In the previous Chapter we have studied the characteristics of only one variable; Example , marks, weights, heights, rainfalls, prices, ages, sales, etc. This type of analysis is called univariate analysis. Sometimes we may be interested to find if there is any relationship between the two variables under study. For Example , the price of the commodity and its sale, height of a father and height of his son, price and demand, yield and rainfall, height and weight and so on. Thus the association of any two variables is known as correlation.

Karl Pearson

Correlation is the statistical analysis which measures and analyses the degree or extent to which two variables fluctuate with reference to each other.

Meaning of Correlation

The term correlation refers to the degree of relationship between two or more variables. If a change in one variable effects a change in the other variable, the variables are said to be correlated.

Types of Correlation

Correlation is classified into many types, but the important are:

- ☆ Positive
- ☆ Negative
- ☆ No

Positive and negative correlation depends upon the direction of change of the variables.

Positive Correlation

If two variables tend to move together in the same direction that is, an increase in the value of one variable is accompanied by an increase in the value of the other variable; or a decrease in the value of one variable is accompanied by a decrease in the value of the other variable, then the correlation is called positive or direct correlation.

Example:

- ☆ The heights and weights of individuals
- ☆ Price and Supply

Rainfall and Yield of crops

- ☆ The income and expenditure

Negative Correlation

If two variables tend to move together in opposite direction so that an increase or decrease in the values of one variable is accompanied by a decrease or increase in the value of the other variable, then the correlation is called negative or inverse correlation.

Example:

- ☆ Price and demand
- ☆ Repayment period and EMI
- ☆ Yield of crops and pest infestation

No Correlation

Two variables are said to be uncorrelated if the change in the value of one variable has no connection with the change in the value of the other variable.

For Example

We should expect zero correlation (no correlation) between weight of a person and the colour of his hair or the height of a person and the colour of his hair.

Simple Correlation

The correlation between two variables is called simple correlation. The correlation in the case of more than two variables is called multiple correlation.

The following are the mathematical methods of correlation coefficient

(i) Scatter diagram

(ii) Karl Pearson's Coefficient of Correlation

Scatter Diagram

Let $(X_1, Y_1), (X_2, Y_2) \ldots (X_N, Y_N)$ be the N pairs of observation of the variables X and Y. If we plot the values of X along x - axis and the corresponding values of Y along y-axis, the diagram so obtained is called a scatter diagram. It gives us an idea of relationship between X and Y. The type of scatter diagram under a simple linear correlation is given below.

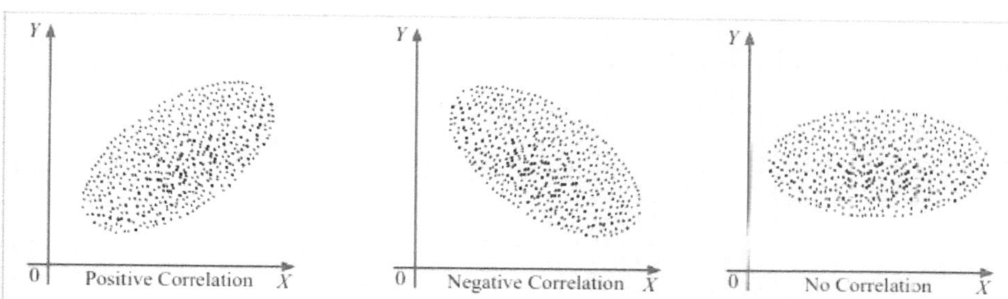

☆ If the plotted points show an upward trend, the correlation will be positive.

☆ If the plotted points show a downward trend, the correlation will be negative.

☆ If the plotted points show no trend the variables are said to be uncorrelated.

Karl Pearson's Correlation Coefficient

Karl Pearson, a great biometrician and statistician, suggested a mathematical method for measuring the magnitude of linear relationship between two variables say *X* and *Y*. *Karl* Pearson's method is the most widely used method in practice and is known as Pearsonian Coefficient of Correlation. It is denoted by the symbol '*r*' and defined as

$$r = \frac{\text{cov}(X,Y)}{\sigma_x \sigma_y}, \text{ where } \text{cov}(X, Y) = \frac{1}{N}\sum(X-\overline{X})(Y-\overline{Y})$$

$$\sigma_x = \sqrt{\frac{1}{N}\sum_{i=1}^{N}(X_i-\overline{X})^2}$$

$$\sigma_y = \sqrt{\frac{1}{N}\sum_{i=1}^{N}(Y_i-\overline{Y})^2}$$

Hence the formula to compute Karl Pearson Correlation coefficient is

$$r = \frac{\frac{1}{N}\sum_{i=1}^{N}(X_i-\overline{X})(Y_i-\overline{Y})}{\sqrt{\frac{1}{N}\sum_{i=1}^{N}(X_i-\overline{X})^2}\sqrt{\frac{1}{N}\sum_{i=1}^{N}(Y_i-\overline{Y})^2}}$$

$$r = \frac{\sum_{i}(X_i-\overline{X})(Y_i-\overline{Y})}{\sqrt{\sum_{i=1}^{N}(X_i-\overline{X})^2}\sqrt{\sum_{i=1}^{N}(Y_i-\overline{Y})^2}}$$

Interpretation of Correlation Coefficient:

Coefficient of correlation lies between –1 and +1. Symbolically, $-1 \leq r \leq +1$

- ☆ When $r = +1$, then there is perfect positive correlation between the variables.
- ☆ When $r = -1$, then there is perfect negative correlation between the variables.
- ☆ When $r = 0$, then there is no relationship between the variables, that is the variables are uncorrelated.

Thus, the coefficient of correlation describes the magnitude and direction of correlation.

Methods of Computing Correlation Coefficient

(i) When Deviations are taken from Mean

Of all the several mathematical methods of measuring correlation, the Karl Pearson's method, popularly known as Pearsonian coefficient of correlation, is most widely used in practice.

$$r = \frac{\sum_i (X_i - \overline{X})(Y_i - \overline{Y})}{\sqrt{\sum_{i=1}^{N} (X_i - \overline{X})^2} \sqrt{\sum_{i=1}^{N} (Y_i - \overline{Y})^2}} = \frac{\sum xy}{\sqrt{\sum x^2 \sum y^2}}$$

Where $x = (X_i - \overline{X})$ and $y = (Y_i - \overline{Y})$; $i = 1, 2 \ldots N$

Steps to Solve the Problems:

(i) Find out the mean of the two series that is $\overline{X}$ and $\overline{Y}$.

(ii) Take deviations of the two series from $\overline{X}$ and $\overline{Y}$ respectively and denoted by x and y.

(iii) Square the deviations and get the total of the respective squares of deviation of x and y respectively and it is denoted by Σx^2 and Σy^2.

(iv) Multiply the deviations of *x* and *y* and get the total and it is denoted by Σxy.

(v) Substitute the values of Σxy, Σx^2 and Σy^2 in the above formula.

9.2 SPEARMAN'S RANK CORRELATION

Spearman's Rank Correlation Coefficient

In 1904, Charles Edward Spearman, a British psychologist found out the method of ascertaining the coefficient of correlation by ranks. This method is based on rank. This measure is useful in dealing with qualitative characteristics, such as intelligence, beauty, morality, character, etc. It cannot be measured quantitatively, as in the case of Pearson's coefficient of correlation.

Rank correlation is applicable only to individual observations. The result we get from this method is only an approximate one, because under ranking method original value are not taken into account. The formula for Spearman's rank correlation which is denoted by ϱ (pronounced as row) is

$$\rho = 1 - \frac{6\sum i_2}{N(N^2 - 1)} \text{ (or) } 1 - \frac{6\sum i_2}{(N^3 - N)}$$

The difference of two ranks = R_i - R_i and N= Number of paired observations.

Rank coefficient of correlation value lies between –1 and +1. Symbolically, $-1 \leq \varrho \leq +1$ When we come across spearman's rank correlation, we may find two types of problem.

(i) When ranks are given

(ii) When ranks are not given

9.3 TEST OF SIGNIFICANCE OF CORRELATION COEFFICIENT

We conduct t-test

H_0 : The Population correlation coefficient ρ = 0

H_1: The Population correlation coefficient ρ ≠ 0

Test Statistic:

$t = \frac{|r|\sqrt{n-2}}{\sqrt{1-r^2}} \sim t_{n-2}$ degrees of freedom with *a* % Level of significance

Inference:

Cal t ≥ tab t, we reject Ho. Otherwise we Accept Ho

9.4 REGRESSION

In regression analysis there are two types of variables. The variable whose value is to be predicted is called dependent variable and the variable which is used for prediction is called independent variable. Regression analysis is a mathematical measure of the average relationship between two or more variables in terms of the original units of the data.

Regression Analysis

Introduction

So far we have studied correlation analysis which measures the direction and strength of the relationship between two variables. Here we can estimate or predict the value of one variable from the given value of the other variable. For instance, price and supply are correlated. We can find out the expected amount of supply for a given price or the required price level for attaining the given amount of supply.

It was first used by British biometrician Sir Francis Galton (1822 -1911), Galton found that the offsprings of abnormally tall or short parents tend to "regress" or "step back" to the average population height. But the term "regression" as now used in Statistics is only a convenient term without having any reference to biometry.

Definition 1

Regression analysis is a mathematical measure of the average relationship between two or more variables in terms of the original units of the data.

1. Dependent and Independent Variables

Definition 2

In regression analysis there are two types of variables. The variable whose value is to be predicted is called dependent variable and the variable which is used for prediction is called independent variable.

Regression helps us to estimate the value of one variable, provided the value of the other variable is given. The statistical method which helps us to estimate the unknown value of one variable from the known value of the related variable is called Regression.

Regression Equations

Regression equations are algebraic expressions of the regression lines. Since there are two regression lines, there are two regression equations. The regression equation of X on Y is used to describe the variation in the values of X for given changes in Y and the regression equation of Y on X is used to describe the variation in the values of Y for given changes in X. Regression equations of (i) X on Y (ii) Y on X and their coefficients in different cases are described as follows.

Case 1: When the actual values are taken

When we deal with actual values of X and Y variables the two regression equations and their respective coefficients are written as follows

(ii) Regression Equation of X on Y;

$$X - \overline{X} = b_{xy}(Y - \overline{Y})$$

where ($\overline{X}$)is the mean of X series,

$\overline{Y}$ is the mean of Y series,

$$b_{xy} = r\frac{\sigma_x}{\sigma_y} = \frac{N\Sigma XY - (\Sigma X)(\Sigma Y)}{N\Sigma Y^2 - (\Sigma Y)^2}$$

is known as the regression coefficient of X on Y, and r is the correlation coefficient between X and Y, σ_x and σ_y are standard deviations of X and Y respectively.

(ii) Regression Equation of Y on X

$$Y - \overline{Y} = b_{yx}(X - \overline{X})$$

where $\overline{X}$ is the mean of X series,

$\overline{Y}$ is the mean of Y series,

$$b_{yx} = r\frac{\sigma y}{\sigma x} = \frac{N\Sigma XY - (\Sigma X)(\Sigma Y)}{N\Sigma Y^2 - (\Sigma Y)^2}$$

is known as the regression coefficient of X on Y, and r is the correlation coefficient between X and Y, σ_x and σ_y are standard deviations of X and Y respectively.

Case 2 Deviations taken from Arithmetic means of X and Y

The calculation can very much be simplified instead of dealing with the actual values of X and Y, we take the deviations of X and Y series from their respective means. In such a case the two regression equations and their respective coefficients are written as follows:

(i) Regression Equation of X on Y:

$$X-\overline{X} = b_{xy}\ (Y-\overline{Y})$$

where ($\overline{X}$) is the mean of X series,

$\overline{Y}$ is the mean of Y series,

$$b_{xy}=r\frac{\sigma x}{\sigma y}=\frac{\Sigma xy}{\Sigma y^2}$$

is known as the regression coefficient of X on Y,

$x=(X-\overline{X})$ and $y=(Y-\overline{Y})$

(ii) Regression Equation of Y on X;

$$Y-\overline{Y} = b_{yx}(X-\overline{X})$$

where $\overline{X}$ is the mean of X series,

$\overline{Y}$ is the mean of Y series,

$$b_{yx}= r\,\frac{\sigma y}{\sigma x}=\frac{\Sigma xy}{\Sigma x^2}$$

is known as the regression coefficient of Y on X,

$x=(X-\overline{X})$ and $y=(Y-\overline{Y})$

Note: Instead of finding out the values of correlation coefficient σ_x, σ_y etc, we can find the value of regression coefficient by calculating $\sum xy$ and $\sum y^2$

Case 3 Deviations taken from Assumed Mean

When actual means of X and Y variables are in fractions the calculations can be simplified by taking the deviations from the assumed means. The regression equations and their coefficients are written as follows:

(i) Regression Equation of Y on X

$$Y-\overline{Y} = b_{yx}(X-\overline{X})$$

where $\overline{X}$ is the mean of X series,

$\overline{Y}$ is the mean of Y series,

$$b_{yx} = r\frac{\sigma y}{\sigma x} = \frac{\Sigma xy}{\Sigma x^2}$$

(ii) Regression Equation of X on Y

$$Y-\overline{Y} = b_{yx}(X-\overline{X})$$

$$b_{yx} = \frac{N\Sigma dxdy-(\Sigma dx)(\Sigma dy)}{N\Sigma dx^2-(\Sigma dx)^2}$$

where dx=X–A, dy=Y–B, A and B are assumed means or arbitrary values are taken from X and Y respectively.

Properties of Regression Coefficients

- ☆ Correlation Coefficient is the geometric mean between the regression coefficients i.e.,

$$r = \mp\sqrt{b_{xy}\ X\ b_{yx}}$$

- ☆ If one of the regression coefficients is greater than unity, the other must be less than unity.
- ☆ Both the regression coefficients are of same sign.

Test the Significance of Regression Coefficient (b) for Y on X:

We conduct t – test

Ho: The Population Regression coefficient $\beta = 0$

H1: The Population Regression coefficient $\beta \neq 0$

Test statistic: $\frac{|b_{yx}|}{SE(b_{yx})}$

$$= \frac{|b_{yx}|\sigma_x^2\sqrt{n-2}}{\sqrt{6_y^2 - b_{yx}{}^2\sigma_x^2}}$$

Regression coefficient (b) for X on Y

Test statistic: $\frac{|b_{yx}|}{SE(b_{yx})}$

$$= \frac{|b_{yx}| \sigma_x^2 \sqrt{n-2}}{\sqrt{ó_y^2 - b_{yx}{}^2 \sigma_x^2}}$$

$$\sigma_x^2 = \frac{\sum x^2}{n} = \frac{\sum (x-\overline{x})^2}{n} \quad \sigma_y^2 = \frac{\sum y^2}{n} = \frac{\sum (y-\overline{y})^2}{n}$$

CHAPTER-10

EXPERIMENTAL DESIGNS

10.1 CONCEPTS OF EXPERIMENTAL DESIGNS

- ☆ Experiment
- ☆ Design of Experiment
- ☆ Treatments
- ☆ Experimental Unit
- ☆ Blocks
- ☆ Yield
- ☆ Experimental Error
- ☆ Principle of Experimental Design
- ☆ Experimental Designs-CRD,RBD and LSD

☆ Experiment

A device or a means of getting an answer to the problem under consideration. Experiments Experiment can be classified into two.

- ☆ Absolute:- designed to calculate certain measures of relationships. *e.g.* Correlation, averages etc.
- ☆ Comparative:- designed to compare the effect of two more characteristics. *e.g.* Comparison of effect of two or medicines, fertilizers etc.

☆ Design of Experiment

The logical construction of the experiment in which the degree of uncertainty with which the inference is drawn may be well defined. The components of experimental design are:

- ☆ Planning of Experiment

- ☆ Obtaining the relevant information from it regarding the Statistical Hypothesis under study Making Statistical Analysis of Data

☆ Treatments

Various object of comparison, in a comparative experiment, are termed as treatments.

e.g. In agricultural experiment, different fertilizers, different verities of crops or different method of cultivation are treatments

☆ Experimental Unit

Smallest division of the experimental material to which we apply the treatments and on which we makes observation on the variable under study the treatments and on which we makes observations on the variable under study is termed as experimental unit.

- ☆ *e.g.* In the agricultural experiments the plot or land is the experimental unit. In medical experiments the unit may be a patient in a hospital, a group of pig etc.

☆ Blocks

In agricultural experiments, most of the times we divide the whole experimental area (field) into relatively homogeneous sub groups or strata These strata, which are more uniform amongst themselves than the field as a whole are known as blocks.

☆ Yield

The measurement of the variable under study on different experimental units are termed as yield.

☆ Experimental Error

Extraneous or random (or chance or non assignable error) variation in the yield due to

- ☆ The inherent variability in the experimental material to which treatment are applied
- ☆ The lack of uniformity in the methodology of conducting the experiment or in other words failure to standardize the experiment technique and
- ☆ The lack of representativeness of the sample to the population under study are called experimental error.

☆ Principle of Experimental Design

The basic principle of the design of experiments are:

- ☆ Replication

- ☆ Randomisation
- ☆ Local Control

Replication

Replication means the execution of an experiment more than once. In other words the replication of the treatments under investigation is known as replication. By replication the experimenter tries to average out as far as possible the effect of uncontrolled factors.

Randomisation

A process of assigning the treatments to the various experimental unit in a purely chance manner. The purpose of randomization is to assure that the source of variation, not controlled in the experiment, operate randomly so that the average effect of any group of unit is zero.

Local Control

The process of reducing the experimental error by dividing the relatively heterogeneous experimental area (field) homogenous blocks is known as Local control.

Types Of Experimental Designs

- ☆ Completely Randomised Design (CRD)
- ☆ Randomised Block Design (RBD)
- ☆ Latin Square Design (LSD)

CRD (Completely Randomised Block Design)

- ☆ Simplest design to use.
- ☆ Design can be used when experimental units are essentially homogeneous.
- ☆ Because of the homogeneity requirement, it may be difficult to use this design for field experiments.
- ☆ The CRD is best suited for experiments with a small number of treatments.
- ☆ Randomization Procedure
- ☆ Treatments are assigned to experimental units completely at random.
- ☆ Every experimental unit has the same probability of receiving any treatment.
- ☆ Randomization is performed using a random number table, computer, program, etc.

Layout

Assign the treatments to the experimental plots randomly

D	A	A	B
B	D	D	A
C	A	C	C
A	A	B	C
D	B	C	A

A, B, C, D are treatments, A - Reapeated $r_1 = 6$, B - Reapeated $r_2 = 4$, C - Reapeated $r_3 = 5$ and D - Reapeated $r_4 = 4$

- ☆ Note that there is no "blocking" of experimental units into replicates.
- ☆ Every experimental unit has the same probability of receiving any treatment.
- ☆ The CRD is the simplest of all the designs based on principle of randomization and replication. In this design treatments are allotted at random to the experimental units over the entire experimental units over the entire experimental materials. Let us suppose that we have k treatments, the ith treatment being replicated ri times i = 1, 2, …, k equal or unequal no. of times. Then the whole experimental material is divided into

$N = \sum_{i=1}^{l} r_i$ for Unequal replication

N= rk for equal replications.

Here, r_i (or) r experimental units and the treatments are distributed completely at random over the units subject to the condition that the ith treatment occurs r_i or r times. The following steps are used to construct the CRD. The following steps are used to construct the CRD

Linear Model

$Yij = \mu + t_i + e_{ij}$ … (1)

μ = overall mean effect

t_i = true effect of the ith treatment

e_{ij} = error term of the jth unit receiving i^{th} treatment

Hypothesis

Ho: All treatments are homogeneous

H_1: All treatments are not homogeneous

Test Statistic

Step 1. Calculate the Grand Total

$$G = \sum_{i=1}^{k} \sum_{j=1}^{ri} Y_{ij} ; N = \sum_{i=1}^{k} r_i$$

Step 2: Calculate the Correction Factor

$$C.F = \frac{G^2}{N}$$

Step 3: Calculate Total Sum of Squares

$$T.S.S = \sum_{i=1}^{k} \sum_{j=1}^{ri} Y_{ij}\,; - D.F.with(N-1)\,d.f.$$

Step 4: Calculate Treatment Sum of Squares

$$T.r.S.S. = \sum_{i=1}^{k} \frac{Y_{i.}^2}{ri} - C.F. \text{ with } (k-1)df$$

Step 5: Calculate Error Sum of Squares

E.S.S = T.S.S –Tr.S.S with (N-k) d.f.

Step 6: ANOVA Table

Source of Variation	d.f.	Sum of Squares (S.S)	Mean Sum of Squares (M.S.S.)	Variance Ratio (V.R.)
Treatments Errors	K-1 N-K	Tr.S.S E.S.S	$TrMSS = \frac{Tr.S.S}{K-1}$ $EMSS = \frac{E.S.S}{N-K}$	$F = \frac{TrMSS}{EMSS}, F_{(K-1\ N-k)}$
Total	N-1			

Inference

Step 7: Compare the calculated F with the critical value of F corresponding to treatment degrees of freedom and error degrees of freedom so that acceptance or rejection of the null hypothesis can be determined.

Step 8: If null hypothesis is rejected that indicates there is significant differences between the different treatments.

Step 9: Calculate C.D. value.

C.D. = SE(d).t_{n-k} (α%)

Here, SE(d) = $\sqrt{2EMSS(\frac{1}{r_i}+\frac{1}{r_j})}$ for unequal number of replications and

SE(d) = $\sqrt{\frac{2EMSS}{r}}$ for equal number of replications

$t_{n-k}(\alpha\%)$ = Table value of t with error degrees of freedom (N-K) at $\alpha\%$

$\alpha\%$ - Level of Significance at 1% or 5% or 10%.

r_i = Number of replications for i^{th} treatment.

r_j = Number of replications for j^{th} treatment.

Advantages of a CRD

- ☆ Its layout is very easy.
- ☆ There is complete flexibility in this design i.e. any number of treatments and replications for each treatment can be tried.
- ☆ Whole experimental material can be utilized in this design.
- ☆ This design yields maximum degrees of freedom for experimental error.
- ☆ The analysis of data is simplest as compared to any other design.
- ☆ Even if some values are missing the analysis can be done.

Disadvantages of a CRD

- ☆ It is difficult to find homogeneous experimental units in all respects and hence CRD is suitable for field experiments as compared to other experimental designs.
- ☆ It is less accurate than other designs.

Applications

- ☆ CRD is most useful in laboratory technique and methodological studies.
- ☆ CRD is also recommended in situations where an appreciable fraction of units is likely tobe destroyed or fail to respond.

Assumptions:

- ☆ All observations are independent
- ☆ Parent population is normal
- ☆ eij~ $N(0,\sigma^2_e)$

10.3 RBD (RANDOMISED BLOCK DESIGN)

- ☆ When the experimental material is heterogeneous, the experimental material is grouped into homogenous sub-groups called blocks. As each block consists of the entire set of treatments a block is equivalent to a replication.
- ☆ If the fertility gradient runs in one direction say from north to south or east to west then the blocks are formed in the opposite direction. Such an

arrangement of grouping the heterogeneous units into homogenous blocks is known as randomized blocks design.

☆ Each block consists of as many experimental units as the number of treatments. The treatments are allocated randomly to the experimental units within each block independently such that each treatment occurs once. The number of blocks is chosen to be equal to the number of replications for the treatments.

Layout

Block 1	Block 2	Block 3
A	D	B
C	A	D
B	C	A
D	B	C

☆ A, B, C, D are the treatments = k = 4

☆ A, B, C, D are repeated equal no. of times = r =3

Analysis

☆ The results of RBD can be arranged in a two way table according to the replications (blocks) and treatments.

☆ There will be r x k observations in total where r stands for number of replications and k for number of treatments.

☆ The data are arranged in a two way table form by representing treatments in rows and replications in columns.

☆ In this design the total variance is divided into three sources of variation viz., between replications, between treatments and error.

☆ The following analysis steps are used to calculate the RBD

Linear Model

$Y_{ij} = \mu + t_i + b_j + e_{ij}$ ... (1)

μ = overall mean effect

t_i = effect of the ith treatment

b_j = effect of the jth block

e_{ij} = error term of the jth block receiving i^{th} treatment

Hypothesis:

Ho:

☆ All treatments are homogeneous

☆ All blocks are homogeneous

H1:

☆ All treatments are not homogeneous

☆ All blocks are not homogeneous

Test Statistic:

Step 1 : Calculate the Grand Total

$$G = \sum_{i=1}^{k} \sum_{j=1}^{ri} Y_{ij} ; N = r * k$$

Step 2 : Calculate the Correction Factor

$$C.F = \frac{G^2}{N}$$

Step 3 : Calculate Total Sum of Squares

$$T.S.S = \sum_{i=1}^{k} \sum_{j=1}^{r} Y_{ij2}; - C.F.with\ (N-1)d.f.$$

Step 4 : Calculate Treatment Sum of Squares

$$Tr.S.S = \sum_{i=1}^{r} \frac{K_{i.}^2}{r} - C.F.with\ (K-1 d.f.)$$

Step 5 : Calculate Block Sum of Squares

$$R.S.S = \sum_{j=1}^{r} \frac{Y_{i.}^2}{r} - C.F.\ with\ (r-1)\ d.f.$$

Step 5 : Calculate Error Sum of Squares

E.S.S = T.S.S –Tr.S.S - B.S.S with (r-1)(k-1) d.f.

Step 6: Anova table

Source of Variation	d.f.	Sum of Squares (S.S)	Mean Sum of Squares (M.S.S)	Variance Ratio (V.R.)
Treatments Blocks Errors	k-1 r-1 (r-1) (k-1)	Tr.S.S B.S.S E.S.S	TrMSS = $\frac{Tr.S.S}{K-1}$ BMSS= $\frac{B.S.S}{r-1}$ EMSS = $\frac{E.S.S}{N-K}$	F = $\frac{TrMSS}{EMSS}, F(k-1), (r-1, k-1)$ F = $\frac{BMSS}{EMSS}, F_{(k-1), (r-1(k-1)}$
Total	N-1			

Inference:

Step 7:

- ☆ Compare the calculated F with the critical value of F corresponding to with appropriate treatment degrees of freedom and error degrees of freedom so that acceptance or rejection of the null hypothesis can be determined.
- ☆ Compare the calculated F with the critical value of F corresponding to block degrees of freedom and error degrees of freedom so that acceptance or rejection of the null hypothesis can be determined.

Step 8: If null hypothesis is rejected that indicates there is significant differences between the different treatments and blocks.

Step 9: Calculate C D value.

C.D. = SE (d).$t_{(k-1)(r-1)}$ (α%);

Here, SE(d) = $\sqrt{\frac{2EMSS}{r}}$ and

$t_{(k-1)(r-1)}$ (α%) = Table value of t with error degrees of freedom ((k-1)(r-1)) at α% l.o.s

α% - Level of Significance 1% or 5%

Advantages

- ☆ The precision is more in RBD.
- ☆ The amount of information obtained in RBD is more as compared to CRD.
- ☆ RBD is more flexible.
- ☆ Statistical analysis is simple and easy.
- ☆ Even if some values are missing, still the analysis can be done by using missing plot technique.

Disadvantages

- ☆ When the number of treatments is increased, the block size will increase.
- ☆ If the block size is large maintaining homogeneity is difficult and hence when more number of treatments is present this design may not be suitable.

Assumptions:

- ☆ All observations are independent
- ☆ Parent population is normal
- ☆ $e_{ij} \sim N(0,\sigma^2_e)$

10.4. LSD (LATIN SQUARE DESIGN)

When the experimental material is divided into rows and columns and the treatments are allocated such that each treatment occurs only once in each row and each column, the design is known as L S D. In LSD the treatments are usually denoted by A B C D etc.

Layout:

For 5 x 5 LSD the arrangements may be

A	B	D	E	C
D	C	E	A	B
C	D	A	B	E
E	A	B	C	D
B	E	A	D	C

- ☆ A, B, C, D, E are treatments
- ☆ Each treatment repeated same no. of times
- ☆ Therefore No. of treatments = No. of replicates = No. of blocks = r

Analysis Steps

Linear Model

The Mathematical model for LSD is

$Y_{ijk} = \mu + r_i + c_j + t_k + e_{ijk}$; i = 1 to r, j = 1 to r and k = 1 to r

r_i is the ith row effect

c_j is the jth column effect

t_k is the kth treatment effect and

e_{ijk} is the error term

Hypothesis

Ho:

- ☆ All treatments are homogeneous
- ☆ All rows are homogeneous
- ☆ All col's are homogeneous

H_1:

- ☆ All treatments are not homogeneous
- ☆ All rows are not homogeneous
- ☆ All col's are not homogeneous

Test Statistic

Step 1. Calculate the Grand Total

$$G=\sum_{i=1}^{k}\sum_{j=1}^{ri}Y_{ij};N=r^2$$

Step 2: Calculate the Correction Factor

$$C.F. = \frac{G^2}{N}$$

Step 3: Calculate Total Sum of Squares

$$T.S.S.=\sum_{i=1}^{k}\sum_{j=1}^{ri}Y_{ij2}-C.F.\ with\ (N-1)\ d.f.$$

Step 4: Calculate Treatment Sum of Squares

$$Tr.S.S=\sum_{i=1}^{r}\frac{Y_{i.}^2}{r}-C.F.\ with\ (r-1)\ d.f.$$

Step 5: Calculate Row Sum of Squares

$$R.S.S=\sum_{j=1}^{r}\frac{Y_{i.}^2}{r}-C.F.\ with\ (r-1)\ d.f.$$

Step 6: Calculate Col. Sum of Squares

$$C.S.S=\sum_{i=1}^{k}\frac{Y_{i.}^2}{r}-C.F.\ with\ (r-1)\ d.f.$$

Step 7: Calculate Error Sum of Squares

E.S.S = T.S.S –Tr.S.S - R.S.S - C.S.S with (r-1)(r-2) d.f.

Step 7: Anova Table

Source of Variation	d.f.	Sum of Squares (S.S)	Mean Sum of Squares (M.S.S)	Variance Ratio (V.R.)
Treatments Rows Cols Error	r-1 r-1 r-1 (r-1)(r-2)	Tr.S.S R.S.S C.S.S E.S.S	TTrMSS = $\frac{Tr.S.S}{K-1}$ RMSS= $\frac{B.S.S}{r-1}$ CMSS= $\frac{C.S.S}{r-1}$ EMSS = $\frac{E.S.S}{(r-1)(r-2)}$	F= $\frac{TrMSS}{EMSS}\sim F_{(r-1,\ (r-1)(r-2))}\ (\alpha\%)$ F = $\frac{RMSS}{EMSS}\sim F_{(r-1,\ (r-1)(r-2))}\ (\alpha\%)$ F = $\frac{CMSS}{EMSS}\sim F_{(r-1,\ (r-1)(r-2))}\ (\alpha\%)$
Total	N-1			

Step 8: Calculate C D value.

C.D. = SE(d). $t_{(r-1)\,(r-2)}\,(\alpha\%)$

Here, SE(d) = $\sqrt{\frac{2EMSS}{r}}$ and

$t_{(r-1)\,(r-2)}\,(\alpha\%)$ = able value of t with error degrees of freedom ((r-1)(r-2)) at $\alpha\%$

$\alpha\%$ - Level of Significance 1% or 5% or 10%

Advantages

- ☆ LSD is more efficient than RBD or CRD. This is because of double grouping that will result in small experimental error.
- ☆ When missing values are present, missing plot technique can be used and analyzed.

Disadvantages

- ☆ This design is not as flexible as RBD or CRD as the number of treatments is limited to the number of rows and columns.
- ☆ LSD is seldom used when the number of treatments is more than 12. LSD is not suitable for treatments less than five.

Assumptions:

- ☆ All observations are independent
- ☆ Parent population is normal
- ☆ $e_{ijk} \sim N(0,\sigma^2_e)$

Index